Flowers in the Snow

Women in the West

SERIES EDITORS

Sarah J. Deutsch
Clark University

Margaret D. Jacobs
New Mexico State University

Charlene L. Porsild
University of Nebraska–Lincoln

Vicki L. Ruiz
Arizona State University

Elliott West
University of Arkansas

FLOWERS
IN THE SNOW

THE LIFE OF ISOBEL WYLIE HUTCHISON

Gwyneth Hoyle

UNIVERSITY OF NEBRASKA PRESS

LINCOLN AND LONDON

*Publication of this book was assisted by a grant from
The Andrew W. Mellon Foundation.*

*Acknowledgments for the use of
copyrighted material appear on p. xvii,
which constitutes
an extension of the copyright page.*

*Library of Congress
Cataloging-in-Publication Data*

*Hoyle, Gwyneth.
Flowers in the snow : the life of
Isobel Wylie Hutchison / Gwyneth Hoyle.
p.   cm.—(Women in the West)
Includes bibliographical
references and index.*
ISBN *0-8032-2403-6
(cloth : alk. paper)*
*1. Hutchison, Isobel Wylie, 1889–1982.
2. Women botanizers—Scotland—Biography.
3. Women travelers—Scotland—Biography.
4. Botanizers—Scotland—Biography.
5. Travelers—Scotland—Biography.
6. Arctic regions—Description and travel.
I. Title.   II. Series.*
QK46.5.B66H69   2001
580'.92—dc21
[B]   00-061592
ISBN *978-0-8032-7344-3
(paper : alk. paper)*

To
the memory of
Medina Lewis
whose devotion to her friend
made this book possible

Contents

Illustrations

Maps

Preface

About twenty years ago, in preparation for a canoe trip eastward on the Thelon River across the Canadian barrens close to the Arctic Circle, I began reading the stories of other women who had ventured north. Excluding those who were sponsored by or connected by marriage to institutions such as the government, the Hudson's Bay Company, or the churches, I found only a handful of women who chose the North as their destination before recreational canoeing there became an established sport. My search of books and articles covering the hundred years beginning in 1840 resulted in an essay titled "Women of Determination."

Not long after the article was published I saw a tiny reference to a book, *North to the Rime-Ringed Sun,* by Isobel Wylie Hutchison. After searching out the book, I read the description of her journey around the north coast of Alaska and her arrival in northern Canada on snowshoes in the depth of winter. I realized that I had missed one of the truly adventurous women of the 1920s and 1930s. My discovery coincided with Isobel Hutchison's death in Scotland in 1982 at age ninety-two. Her obituary in the *Scottish Geographical Magazine* briefly listed her achievements and honors and noted her long connection with the Royal Scottish Geographical Society.

It was more than ten years before I was free to make amends for omitting this vital northern traveler from my article. When I contacted the Society, Isobel Hutchison was remembered for her donation of a large collection of photographs to its archives, but those who answered my request for information did not remember her personally. They were, however, able to suggest Lady Kathleen Dalyell, the wife of a Scottish parliamentarian, who had known her in her late years. Through Lady Dalyell I learned of the 1987 memorial exhibition of Hutchison material mounted at the National Library of Scotland in Edinburgh, where the bulk of her papers are held.

With that contact, I made a visit to Edinburgh for a preliminary look at the archival holdings. Those who remembered Isobel Hutchison urged me to proceed without delay to tell the story of her life—she was in danger of being forgotten. From those tentative beginnings, a series of happy coincidences led me to wonder occasionally if my path was being directed.

Among the first people I met were Katie and Wilson Marshall, who live at Carlowrie, the house where Isobel Hutchison was born and died, the last of her line. Although they had not known her personally, the Marshalls knew of her accomplishments and had a collection of papers relating to the Hutchison family. In addition, they put me in touch with Sir Peter Hutchison, a second cousin, and with neighbors and others who had known her.

Sir Peter, who has since died, graciously lent me his file of material about his adventurous cousin. Among the papers were many letters written to Sir Peter by Isobel's closest friend, Medina Lewis, who was with her frequently before her death and who sorted her papers for preservation in the archives. One of the letters was written from the home of her niece, Katrina Burnett, whom I found to be still at the same address fifteen years later. Mrs. Burnett supplied me with many of the family details about her aunt and helped me to appreciate the deep friendship between the two women.

On her northern travels, Isobel Hutchison collected plants for the Royal Botanic Gardens at Kew and for the British Museum, and both institutions supplied me with files of her correspondence. At the Royal Scottish Museum in Edinburgh, the curator, Briony Crozier, showed me the artifacts Isobel had purchased for or donated to the collection. In the herbarium of the Royal Botanic Garden in Edinburgh I saw some of the plants she had collected as well as one of the small plant presses she had used. Artifacts and papers were seen at the Museum of Archaeology and Anthropology and at the Scott Polar Research Institute in Cambridge. I found more correspondence in the archive of the Royal Geographical Society in London.

The thick files of letters in the National Library archive in Edinburgh attest to Isobel as a prolific letter writer. Unfortunately, her letters to her family or to her friend, Medina Lewis, were not saved. It is possible that she wrote few on her northern trips because for the most part she was beyond regions where there was a regular postal service. On this side of the Atlantic, I obtained the file of letters exchanged between her and the explorer Vilhjalmur Stefansson, from the Stefansson Library of Dartmouth College in New Hampshire.

At Trent University, staff members were ever willing to assist in my project. Linda Matthews in the Reference Department of the library found obscure addresses for me in Britain and Denmark. Winnie Janzen translated Danish documents from her native tongue. The Interlibrary Loan Department obtained numerous books and articles for me. Several professors were generous with their time. Roger Jones of the Botany Department gave me insight into the importance of plant collecting. Suzanne Bailey, a specialist in the literature of travel, was always ready to discuss aspects of Isobel Hutchison's life and travel, and Joan Sangster of Women's Studies answered my many questions. John Jennings gave valuable assistance in collecting photographs. John Wadland of the Frost Centre for Canadian Studies and Native Studies was always encouraging.

Colleagues at universities beyond Trent were supportive. Kenneth Coates from the University of New Brunswick, Sherrill Grace from the University of British Columbia, and Ian MacLaren from the University of Alberta all provided helpful comments. Richard Winslow III, friend and librarian in New Hampshire, sent me many useful articles from a variety of sources.

Throughout the search for Isobel Hutchison my husband, Alex, was an invaluable participant, accompanying me on research trips to Scotland, reading and commenting on what I had written, and sharing countless meals with our frequent phantom guest, IWH.

Acknowledgments

I thank the following for permission to use copyrighted material:

The Trustees of the National Library of Scotland, Edinburgh

Mary-Grace, Lady Hutchison, MBE, for Hutchison family papers

Gresham Publishing Company, New Lanark, Scotland, for permission to use quotations and photographs from *North to the Rime-Ringed Sun, Arctic Nights' Entertainments, Stepping Stones from Alaska to Asia*, and two poems from *Lyrics from Greenland*

The Archives, Royal Botanic Gardens, Kew

Hudson's Bay Company Archives, Provincial Archives of Manitoba, Winnipeg

Royal Scottish Geographical Society, Glasgow

Trustees of the Natural History Museum of the British Museum, London

Scott Polar Research Institute, Cambridge University

Dartmouth College Library, Hanover, New Hampshire

National Archives of Canada, Ottawa

University of Washington Press, Seattle, for permission to quote from *Journal of an Aleutian Year* by Ethel Ross Oliver

I also thank James S. Baird for use of a quotation from his letter, Katie Marshall, for permission to photograph her home, and Agnes (Nessie) McGowan, for the photograph from her wall.

Flowers in the Snow

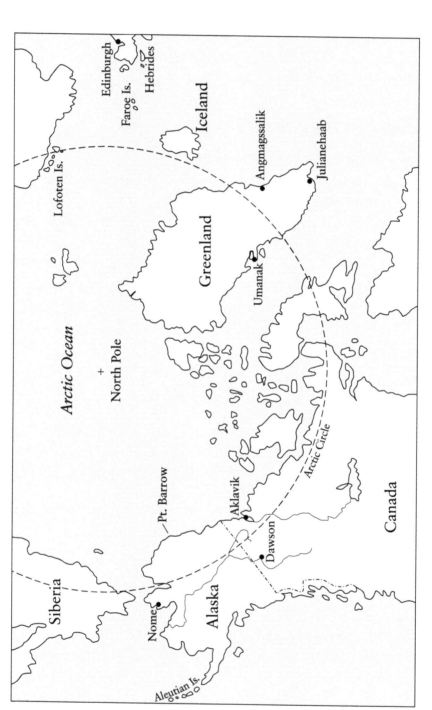

Arctic Circle destinations

Introduction

On the tundra above the bleak Arctic coast of Alaska a woman knelt to admire a solitary yellow poppy thrusting up through the snow, its color of distilled sunshine filling her with optimism. Like the flower still blooming in October, she had come late and alone to Arctic travel. Protecting the delicate bloom was a thin sheath of ice, like the cool, polished, ladylike exterior that shielded the private nature of an artist and a poet. As she watched the poppy blowing in the wind on its slender stem and yet firmly rooted, she recalled how she had been buffeted by the conflicting advice of everyone she had met on her travels yet held fast to her original purpose. The Arctic poppy, with its contradictions of delicacy and toughness, is a fitting symbol for the woman who found it.

The woman is Isobel Hutchison, and the book in which she tells of finding the Arctic poppy is *North to the Rime-Ringed Sun.* Though the book is written without hyperbole and acknowledges few difficulties, the adventure she recounts is one of breathtaking boldness. This self-effacing modesty has helped to make her name less recognized than most in the select company of adventurous women traveling in the early years of the twentieth century.

In 1949 the ancient University of St. Andrews in Scotland awarded an honorary doctorate of laws to this quiet, unassuming Scottish lady, who was surprised to be singled out for the honor. The citation at the granting of this degree gave a graceful summary of the life and work of Isobel Hutchison:

[A] scientist by training, a poet at heart, [she] has braved the lonely icy wastes of Greenland and Alaska, the mist and fog of the Aleutian Islands, and the untrodden spaces of Canada not only, we believe, to collect plants, but also, we surmise, to satisfy the restless surging of that indomitable spirit which defies hazard, danger and discomfort, and is the source of all great human achievement. Journeyings worthy of romantic saga, contributions to the rich collections of rare plants gracing our botanical gardens, books swelling the exciting literature on Arctic travel, these are signal achievements. They are enriched by a mastery of six strange tongues, and novels and poems written in her own.[1]

In a single decade Isobel Hutchison ventured progressively westward from Norway's Lofoten Islands along the Arctic Circle to the Alaskan edge of the Western Hemisphere. Her destinations were not conceived as part of a great travel plan; like the plants and flowers she sought, they just grew. She made four major northern journeys between 1927 and 1936, two to Greenland, another along the northern coast of Alaska and into Arctic Canada, and one exploring the full extent of the Aleutian and Pribilof Islands. As the director of the Scott Polar Research Institute at Cambridge University wrote at the time of her death in 1982, it was the unexpectedness of her travels and the determination with which she pursued her goals that caught one's attention.

Born into a wealthy home, Isobel readily accepted "hard" travel as the condition to achieve her goals. The interest on a small legacy she received from her father's estate when she was twenty-one provided just enough financial independence to permit her to follow her own paths in her own way. By endowing all five children equally, Thomas Hutchison was uncharacteristic of Victorian parents, who tended to favor sons over daughters.[2] His widow held firmly to the Victorian belief that only sons needed higher education, but she promoted conventional travel on the European continent for all of her family. When her middle child, Isobel, began her adventurous expeditions, her mother considered the destinations outlandish; but since Isobel was using her own resources, there was little her mother could do to prevent it.

With a strong streak of independence and quiet determination, Isobel Hutchison found the freedom her spirit craved in the world of nature. Intensely spiritual, she perceived the face of God in distant horizons, in the sea in all its moods, and in the infinite variety of plants and flowers. Encouraged by her father, she had collected, pressed, and mounted flowers from childhood, and this passion for botany would unlock the gate to the North.

Growing up in Scotland, she early discovered the freedom to be found in long, solitary walks. In her middle thirties, in 1924, she walked the length of the Outer Hebrides, and the following year she made a solo trek across Iceland, after being told by the local people that it could not be done. Looking ever northward, Greenland seemed the next logical step. But Greenland was forbidden territory. To protect the aboriginal people, imperial Denmark denied entry to all travelers who could not show valid reasons to go there, such as the intention to carry out scientific research or exploration. Undaunted, Isobel applied to the authorities in Denmark to make a private study of Greenland's plant life on a summer trip. Then, having been captivated by the grandeur of the country, she applied for permission to return for a whole year and live in a native community north of the Arctic Circle, where she would collect plants for the Royal Horticultural Society.

Without formal university training as a botanist, Isobel Hutchison was a plant collector in the mold of the self-taught nineteenth-century naturalists Alfred Wallace and Thomas Huxley. As was said of the great botanist Sir Joseph Hooker, a director of the Royal Botanic Gardens at Kew, she "did not so much learn botany as grow up in it."[3] In a similar fashion, Isobel absorbed a love of plants and nature from her father, who was secure enough financially to retire at the time of her birth in 1889 and devote himself to his family, his estate, and his gardens. Her enthusiasm for the plants of the Scottish Highlands transferred naturally to the alpine species she found in the stark, uncluttered landscape of the Arctic.

In the Victorian era, collecting and pressing plants and flowers were considered suitable occupations for women and played into the rising importance of taxonomy, the naming and classifying of species. Women living in the far reaches of the British Empire collected plants for the Royal Botanic Gardens at Kew, often from southern or tropical regions. Sharing the enthusiasm and passion of women settlers on the frontiers of Australia and Canada who explored and found new plants regardless of the obstacles to be overcome, Isobel Hutchison preferred to work in open and mountainous terrain.[4]

Her successful work for the Royal Horticultural Society led to future plant-collecting assignments for the Royal Botanic Gardens at Kew and the

British Museum. Cambridge University and the Royal Scottish Museum in Edinburgh were also interested in having any native artifacts she could procure. Accreditation by such world-famous institutions, though it did not make travel conditions easier, helped her to negotiate bureaucratic hurdles.

Painfully shy as a child, lacking self-confidence as a young adult, and leading a sheltered life in a conventional Victorian home until she was nearly thirty, Isobel expanded and blossomed in personality with each new adventure of independent northern travel. She was like a flower whose bud remains tightly closed until the right circumstances cause it to open.

The early years of such a private person are sketched in the diaries she faithfully kept from adolescence, diaries that recorded facts and events but rarely emotions. Deep feelings surfaced in *Original Companions*, a novel that Isobel Hutchison wrote in her early thirties. The novel served as a catharsis at a time of deep emotional crisis, and she described it in her diary as truth. In the catalog of the memorial exhibition held at the National Library of Scotland shortly after her death and arranged by her closest friend, Medina Lewis, the novel is designated as autobiographical. With a knowledge of her background and home territory, it is not difficult for a discerning reader to separate fiction from truth in the story and to see how much she has revealed of herself.

Writing the novel seemed to release Isobel from the emotional bonds that held her captive. Tentatively and then with growing assurance she began to journey toward the beckoning horizons. Poetry and travel articles flowed from her pen as she used her natural gift to augment her modest income. As the trips became longer and more adventurous, her correspondence with an ever-widening circle of friends grew. Ample files of letters held in archives on both sides of the Atlantic add greatly to the understanding of her personality.

The books that resulted from her travels, *On Greenland's Closed Shore, North to the Rime-Ringed Sun, Arctic Nights' Entertainments,* and *Stepping Stones from Alaska to Asia,* help to complete the portrait of this fascinating lady. Isobel Hutchison's books consistently underplayed the difficulties of her journeys. After a voyage on storm-tossed or icebound seas, she would arrive in a small community, sometimes where no English was spoken, without friends or letters of introduction. She was always willing to accept strange food and any available accommodations, and her charm readily made friends for her. She did not pass judgment, and there are few if any adverse comments on the people she met. Unlike the Victorian traveler Isabella Bird, who could be judgmental, acerbic and even peevish in her descriptions, she was aware that her books might be read by those who had given her hospitality and was careful not to offend them. Although her accounts are true and accurate, she emphasized the positive aspects and barely hinted at the negative. Rather than sensationalizing the problems she faced, she preferred to share with readers her pleasure in the northern landscape and northern people.

After the publication of each book, Isobel Hutchison was much in demand as a lecturer—at the Royal Geographical Society in London, the Scott Polar Research Institute of Cambridge University, and the Royal Scottish Geographical Society, as well as in village halls throughout Scotland and parts of England. She also gave short radio talks for the British Broadcasting Corporation. In the 1920s and 1930s, before the days of mass tourism and lavish commercial travelogues, before television or the Internet could take the armchair traveler to myriad destinations by a turn of the dial or a click of a mouse, the hunger for information about distant lands was fed only by books and illustrated lectures. From her extensive travels in the North, Isobel Hutchison brought back information and tales that were available from few other sources. During her lifetime she gave more than five hundred illustrated lectures.

What magnetic quality of the North continually drew her back? On her first visit to Greenland, Isobel met the famous explorer Knud Rasmussen and formed a deep friendship with him, later receiving an inscribed copy of his book, *Across Arctic America.* An ethnologist and legendary northern traveler, only a few years earlier Rasmussen had made a three-year trek from Greenland to Nome on the western coast of Alaska in an attempt to prove the kinship of all northern natives through their customs and folktales. Although much of Isobel Hutchison's travel after Greenland depended on happenstance, her inspiration came from Rasmussen's example. In the course of her northern travel, she made contact with native people whenever possible, always trying to converse with them in her limited Greenlandic language. At the same time, she observed the effects on their way of life of three different jurisdictions—Danish, American, and Canadian—as well as the remnants of Russian influence in the Aleutians.

In the nineteenth century northern exploration and travel were a male preserve, dominated by naval officers, whalers, and sportsmen. In the early years of the twentieth century, the public imagination was captured by polar exploration, with Robert Peary reaching the North Pole in 1909 and Roald Amundsen the South Pole in 1911. After the hiatus caused by the First World War, explorers returned to the North when Oxford and Cambridge Universities mounted large and small expeditions in the 1920s. Travel to the North became increasingly commonplace in the 1930s as many European countries and several other British and American universities sent expeditions to Iceland, Greenland, Spitsbergen, and the Canadian Arctic.[5] The accounts of their explorations were published in academic and scientific journals but rarely in popular form accessible to the general public. Beginning her northern travels in the 1920s, Isobel Hutchison was in the vanguard of these explorers, but unlike them she was never part of an organized or sponsored group.

In the 1920s and 1930s it was unusual for women to go north beyond the goldfields of Alaska and the Yukon except for reasons of duty, and it was certainly not something to be undertaken lightly by a single woman traveling alone. In this period she falls between the much-studied Victorians and the adventurous women of the present, in that small group of largely forgotten travelers of the early twentieth century. Born of Victorian parents and raised in the Edwardian era, these women experienced many of the restrictions that had hobbled their earlier sisters, but they also began to taste the freedom of their later counterparts.

By making light of her difficulties, Isobel did nothing to increase her reputation as an adventurous traveler, as she might have done had she exaggerated her accomplishments. Concerning her boldest travel, eastward by dog team along the Arctic coast of Alaska into Canada as winter approached, she claimed only two distinctions for herself: she was the first non-native woman to enter Canada at Demarcation Point, and she was the only woman to board the derelict Hudson's Bay Company ship *Baychimo*, imprisoned in the ice floes of the Arctic Ocean.

Her success as a traveler, combined with her essential modesty, has caused Isobel Hutchison's name to be almost forgotten. The public imagination is captured by tragedy and failure. The death of Scott and his companions in Antarctica in 1912 soon after reaching the Pole is remembered over

Shackleton's heroic escape, with no loss of life, from the crushing ice floes of the Antarctic only a few years later. Among the northern travels of women, the canoe journey of Mina Hubbard through Ungava in 1905 is dramatized and made more memorable by the tragic death of her husband on a failed attempt at the same journey. By the same token, Florence Tasker's successful canoe traverse of northern Ungava with her husband in 1906 is barely recalled.[6] Isobel Hutchison's lack of life-threatening accidents and tragedies has relegated her to the far corners of exploration history.

Tragedy was not unknown to Isobel, but she had put it behind her when she began her travels. Overcoming the early loss of three loved family members was a vital factor in strengthening the indomitable spirit that carried her through times of loneliness, discomfort, and danger on her journeys. Her writings, too, are less about adventure than about the magnificence of nature in the North and about the people who live there.

It is a paradox of her personality that although she cherished privacy she accepted and thrived in the public role that came with her lecturing. Her personality encompassed other paradoxes as well: her love of unconventional travel outside the conservative framework she lived in; her childlike sense of wonder combined with her sophisticated understanding of world events; her joy in physical activity as an adjunct to her purely cerebral energy; her fanciful imagination housed in a practical mind.

She was something of a Renaissance woman, her achievements many and various. In addition to producing books of travel, poetry, lectures, journalism, and paintings, she collected thousands of specimens of Arctic plants, which are now held in several British herbaria and continue to add to plant knowledge throughout the world. Her collections of artifacts from Greenland and Alaska were made at a crucial time, just before authentic items were being replaced by native handiwork done for the tourist trade, and they are treasured both at Cambridge University's Museum of Archaeology and Anthropology and at the Royal Scottish Museum in Edinburgh.

For her achievements Isobel Hutchison received many awards, including the Mungo Park Medal of the Royal Scottish Geographical Society and the Freedom Medal from Denmark in addition to the one she considered the pinnacle of her career: the honorary doctorate of laws from the University of St. Andrews.

The story of her life is the record of an accomplished amateur whose unfailing courtesy and good humor, wide interests, and fearlessness carried her across the top of the world, developing friendships at each stop along the way. The extraordinary life of Isobel Hutchison is a testament to what can be achieved by an ordinary person who has the courage to be herself and the determination to pursue her dreams to the ends of the earth. A life to be remembered and celebrated!

I

Carlowrie

Half hidden among the ancient trees, the tower and turrets of Carlowrie appear the very model of a Scottish baronial castle. This is the house where Isobel Hutchison was born and where she died ninety-two years later. It was the haven she returned to after months spent in travel to destinations far removed from the European world she knew. It was the home where her roots were deep and firm.

Carlowrie was built in 1852 by Isobel's grandfather, Thomas Hutchison. The son of a flesher, or cattle breeder, he was born in 1796 at Kinghorn in Fifeshire, but his destiny lay across the Firth of Forth in the town of Leith, on the outskirts of Edinburgh. There he worked for a firm of wine and spirit merchants, George Young and Company, owners of the Grange Distillery. Astute and quick to learn, by age thirty he had founded his own wholesale wine business, T. Hutchison and Company. The times were favorable for the wine trade: the Industrial Revolution created a wealthy middle class that had the leisure to enjoy the pleasures of the table and could afford the luxury of good claret.[1]

Well established in the wine trade, in 1832 Thomas Hutchison married Jean Wylie. Her grandfather, William Wylie, was a farmer near Kincardine, then in Perthshire, close to the River Forth. The sons of William Wylie followed diverse paths. While Jean Wylie's father, Robert, remained on the family farm, her uncle James became the physician to three Russian czars, Paul, Alexander, and Nicholas. He was one of the founders of the Medical Academy of St. Petersburg and Moscow and its president for thirty years. At the request of Czar Alexander, in 1814 he was knighted in

England by the prince regent, acting on behalf of King George III. He died many years later at St. Petersburg, leaving half his fortune to the czar. Jean Wylie's youngest uncle, Walter, was a sea captain, sailing from the River Forth to Baltic ports to sell pit props to the Russians. Among Jean Wylie's seven brothers and sisters, four brothers followed their Uncle Walter to sea, including one who died in Moscow and a brother, as well as a sister, who married in Canada.[2]

Thomas Hutchison and his wife had a country home, Glendevon House, in the Ochil Hills of Perthshire, where their five children were born. They also owned a town house, the Hermitage, near the golf course in Leith. As his business prospered Thomas Hutchison became active in local politics, being elected provost of Leith in 1845. He promoted improvements to the harbor and the rail service as well as holding directorships in important Scottish financial institutions.[3]

Having amassed a considerable fortune by 1850, Thomas Hutchison bought the estate of Carlowrie, near the village of Kirkliston in West Lothian, closer to his business interests in Edinburgh. The existing house was pulled down, and Edinburgh architect David Rhind was engaged to build a new house in the Scottish mid-nineteenth-century baronial style. The architect's specifications called for the finest stone and woodwork, and the final cost of £33,000 represented a very large sum at the time.[4]

Approached through an avenue of sycamore trees, the formal entrance to Carlowrie is set on a raised terrace. The entire edifice is surmounted by a round tower with an arched balustrade, and turrets with candle-snuffer roofs complete the baronial effect. A large conservatory filled with rare and tropical plants is off the formal drawing room to the east of the entrance; it was originally much larger, extending almost the width of the lawn. The house is set in spacious grounds, with walled gardens hiding the greenhouses, kitchen gardens, and stables. A second avenue of trees leads westward past the paddock to the lodge beside the now-quiet country road. A three-hundred-acre tenant farm on the other side of the estate, with its sturdy eighteenth-century farmhouse and outbuildings, supplied milk and produce for the big house. When Carlowrie was built facing south across the River Almond, the setting was peaceful in spite of being close to one of the main roads to Edinburgh. From the open tower of the house one can look south to the Pentland Hills stretching off into the distance and north toward the Firth of Forth.

The house was two years in completion, and Thomas Hutchison died before it was ready to occupy. His widow lived there with her four sons until 1863, when the eldest, Robert, married the daughter of the local Presbyterian minister. At Carlowrie Robert enjoyed the role of country gentleman, with a passionate enthusiasm for archaeology, rural economy, and arboriculture. He published several essays and papers on rural economy for private circulation, and he was made a fellow of the Royal Society of Edinburgh as well as belonging to other Scottish learned societies. He and his wife had seven children, two of whom were later knighted for their accomplishments.[5] Unfortunately Robert Hutchison, despite his scholarly achievements, had no interest in the family wine business, and by 1888 his extravagant lifestyle had run him deeply into debt.

Robert's younger brother, Thomas, having followed his father in the wine trade, spent a number of years in India expanding the business. When he returned to Scotland at about age forty, he married his second cousin, Jeanie Wylie, the granddaughter of his mother's seafaring uncle, Walter. Her father was a farmer and maltster from Parkhead, near Alloa, and her mother was a Younger, whose family founded a famous Alloa brewery. For the first years of their marriage Thomas Hutchison and his wife lived in his father's home, Glendevon House, where two children, Nita and Walter, were born.

To Thomas fell the burden of saving the family's reputation and business. In 1888 he paid off his brother's debts and took over the ownership of Carlowrie, where, on May 30, 1889, a third child, Isobel Wylie, was born. Hilda and Frank followed—five children over the span of twelve years.

The wholesale wine trade was flourishing. Soon Thomas Hutchison could afford to devote his time to his family and his estate. When it was necessary to attend to business matters, the frequent trains from nearby Ratho station would take him to Edinburgh in less than twenty minutes. He had a passionate interest in the gardens, and in his daily journal he recorded his observations of the weather and details of the plantings. A quiet, reserved man, he imparted his love of nature and horticulture to his children by example. By the time his fifth child was born, when he was in his late fifties, his garden, his library, and his family were the center of his life.

Carlowrie was a world unto itself, a halcyon domain where "it seemed always afternoon." A dozen servants were on hand to respond to the double row of bells hanging in the hall beside the kitchen. The gardener and his wife lived in the lodge at the end of the avenue of trees, and the tenant farmer was the nearest neighbor on the other side of the estate. A mile up the country road was the village of Kirkliston, and each Sunday the Hutchison family walked to the twelfth-century kirk where Thomas Hutchison was an elder and took their places in the front pew of the gallery.

A resident governess took care of the schooling, sometimes assisted by a fräulein to teach the children German. After lessons there was plenty of scope for five active youngsters, with croquet, tennis, archery on the lawns, skating in winter, bicycling, and games of hide-and-seek at any time. The days were never long enough, filled with hikes in the country, picnics, and visits to and from the Wylie relatives on the other side of the River Forth. There were trees to climb, where one could disappear and read in secret leafy bowers. The children also loved to write and produce plays, using the spacious first landing of the great staircase as a stage. The servants, seated in the formal entrance hall, provided the audience for the first performance, and relatives were invited for subsequent evenings of entertainment.[6]

Before she was ten, Isobel had her own garden plot near the greenhouses where, with her father's encouragement, she built a cold frame, planted seeds, and recorded the growth of her plants. Her older brother, Walter, was a keen photographer and taught her to develop and print film in the darkroom. Lessons in Scottish dancing and physical sports like running and jumping were as much a part of her life as the more ladylike pursuits of embroidery, crocheting, painting, and—significantly—pressing flowers. Idle moments were few, and she cherished time alone to absorb the beauty of the world around her, to think about the mysteries of life, and to attempt to frame her thoughts into poems. For all her love of sports and physical activity, Isobel was a shy and introspective child. Surrounded by a close and loving family, with no need for outside friendships, her life at Carlowrie was as perfect as she could wish it to be.

With the dawning of a new century, the tranquil family life of Carlowrie was shattered. In April 1900 Thomas Hutchison caught a chill that rapidly turned to pneumonia. Within three days the husband and father, the center of their golden world whose constant presence was the vital spark, was gone.

Isobel's father died just before her eleventh birthday. In her only novel, written more than twenty years later and describing circumstances that closely matched her own, she reproduced the powerful but inexpressible feelings of a child facing the sudden death of a beloved parent. Her despair and disbelief mingled with indignation at the unrestrained emotions of the adults around her. Too proud to cry in front of the servants, she felt her unshed tears enter like iron into her soul, and with the anguish of a wounded wild creature she hid her grief in solitude, only to have it surface again in years to come.[7]

Although his death was unexpected, with his usual attention to detail Thomas Hutchison had left his affairs in perfect order. All financial matters were arranged to be administered by an Edinburgh law firm as a strict trust, which continued until the death of Isobel, the last surviving member of the family. A nephew, the third Thomas Hutchison, son of the improvident older brother Robert, was already working in the family business and became managing director when it later amalgamated with J. C. Thompson and Company, wine merchants.[8]

The arrangements Thomas Hutchison had made ensured that the family could continue to live in its accustomed style, with adequate indoor and outdoor staff. While a lawyer with the title of factor oversaw the management of the estate at Carlowrie, Mrs. Hutchison—Mama as she was known by her children—took over the direction of the family.

Sixteen years younger than her husband, Mrs. Hutchison was small and dark haired with a brisk and capable manner. At school she had been one of the brightest girls in her class, able to hold her own in the subjects taught to members of both sexes.[9] As a matron, she directed the large household of servants with a firm hand. After the initial shock of her husband's death she took full charge of bringing up the five children, ranging in age from sixteen to four. Whereas her husband's interests had centered on the estate and the gardens, Mrs. Hutchison's world was that of the conventional Victorian lady. Now a widow, she spent her leisure hours entertaining visitors or being driven in the brougham to leave calling cards with suitable neighbors. She had a close relationship with her mother and aunts in Alloa and with her sisters, all living within easy journeys of Carlowrie. She and her sisters suffered from congenital deafness, and in later years Mrs. Hutchison was afflicted by tinnitus, or ringing in her ears.[10]

Writers often reveal more of their true feelings in their imaginative fiction than through straight autobiography.[11] In *Original Companions* Isobel Hutchison described her mother as full of common sense with a wholesome gift of humor but lacking in imagination, while her father was lavishly generous in a secretive way. In her diaries her relationship with her mother can be perceived as correct and conventional. Whether deafness was the barrier or Mrs. Hutchison's formal manner, Isobel's affection for her mother appeared to lack the depth of feeling she had for her introspective, nature-loving father.

Isobel was a faithful diarist. From a tentative beginning when she was ten years old, the events of most of the next eighty years were chronicled in commercial diaries about three inches by four or even smaller, her open, rounded script filling every part of each tiny page. The diaries, ten or more to an archival box, are housed in the Archives of the National Library of Scotland in Edinburgh, each page giving a terse summary of the day's activities. There is little space for more than facts. Some particularly momentous years are inscribed in greater detail in hard-cover notebooks, and in these she recorded not only daily events but thoughts and feelings.

Most children find keeping a daily diary a tedious exercise; they often begin boldly at the start of each new year, but the entries peter out in a few months. In her first diary in 1899 at the age of ten, Isobel made sporadic entries until May, when she received a new bicycle for her birthday. There are scattered entries for the years 1901 and 1902, but at age fourteen, in 1903, she had developed the discipline to maintain daily entries that lasted to the end of her life. Significantly, the diary for 1900, the year her father died, and those for the years when she suffered other tragic losses are missing from the collection. It seems that any reminders of those years were so painful they must be expunged from the record. Although her diaries reveal little of her inner self, their existence shows that from an early age she was self-disciplined, regular in her habits, and diligent. The absence of diaries for specific tragic years is the first indication that Isobel's feelings ran so deep that she hid them even from herself.

The early diaries show Isobel as an active teenager, something of a tomboy who delighted in bicycling, golf, cricket, high jumping, and running. While the diaries reported mainly the life within the enclosed world of Carlowrie, distant events such as the death of Queen Victoria or the Russian-Japanese War were mentioned. More regularly recorded were the books Isobel read—Dickens, Thackeray, and many of Scott's Waverley novels, as well as popular novels of the time and the *Girl's Own Paper.* Equally important in her life was writing: plays to be performed for invited audiences, regular contributions to the family magazine, the *Scribbler,* and always poems.[12]

Begun in 1903 with Nita as the first editor-in-chief, the *Scribbler* was produced every two months and contained typed articles, stories, poems, plays, and nature notes by the three sisters and some by outside contributors. For a few years Walter provided photographs and articles on photography. The artwork, both inside the magazine and on each unique cover, showed a high degree of competence and originality. The subscribers—friends and extended family members—received the one copy of each issue in turn, crossing off their names and posting it to the next on the list. From the first issue it was clear that the Hutchison sisters, Nita and Isobel in particular, were serious about writing, even in their humorous pieces. In later years when Nita had left home and Hilda was studying abroad they continued to submit stories and poems and, in Hilda's case, musical compositions. Isobel, still at home, became the editor-in-chief, writing editorials with dry wit and illustrating stories with clever cartoons and caricatures. Over the course of eight years, twenty-three issues were produced.

Before her father's death it had been arranged that the eldest daughter, Nita, would attend a boarding school, Calecote Towers, in Hertfordshire. Walter, the older son, became a day pupil at Fettes, a prestigious Edinburgh school. These two had been Isobel's constant companions from her earliest childhood. Her younger sister Hilda now began to take Nita's place, and Walter became even more important to her during his free time from school. Frank, the baby of the family, was seven years younger than Isobel, and it would be a few years before he was part of the close circle.

Walter completed his schooling at Fettes and passed the preliminary examination to study chartered accountancy. In the summer of 1904 Mrs. Hutchison took Nita and Walter to France for a holiday, and Walter remained until Christmas at the University of Grenoble. Miss Whitelaw, their longtime governess, retired from teaching to look after her mother, and in the same year Isobel, Hilda, and Frank began attending private school in Edinburgh—the girls at Miss Gamgee's, which later became Rothesay House, and Frank at Miss Menzies's.[13] They traveled into the city each day by train from Ratho station and walked from Haymarket station to the schools in the west end of Edinburgh. Rothesay House, a small school occupying two houses on Rothesay Terrace, taught a curriculum designed for young ladies expecting to live privileged lives at home, following the patterns of their mothers. Isobel often stood at the top of her class, and botany was among the subjects she excelled in.[14] Daily commuting distanced the Hutchison girls from their classmates, who either lived in Edinburgh or boarded at the school. Always more at ease with older people and young children, Isobel might have seemed aloof, but with plants, trees, and books as her companions she was never lonely.

In Scotland, influenced by the Protestant Reformation, universal education for both sexes was the accepted tradition, and a few women had even attended university late in the eighteenth century. University education was stressed for all teachers, and teaching was considered an honorable profession.[15] Mrs. Hutchison, however, from a family that did not see the need of higher education for girls, accepted that view without question. Though her oldest daughter Nita showed great literary and artistic promise at her school in England and the headmistress urged her to attend university, Mrs. Hutchison thought it unnecessary. Marriage was the proper course for girls.[16]

Although Thomas Hutchison was a successful merchant and prosperous landowner, he was not "landed gentry" or even "county" in the subtle rankings of British aristocracy. Carlowrie, designated on the map as a castle, was in reality a baronial mansion, set in spacious grounds. It was called a castle to distinguish it from Carlowrie farm, part of the extensive estate of Lord Rosebery, the prime minister who followed Gladstone, which bordered Carlowrie to the northeast. The children at Carlowrie were allowed to speak with the naturally soft lowland Scots accent, whereas the governess at Dalmeny House, the Rosebery mansion, taught her charges to speak with the proper upper-class English accent.[17] The daughters at Carlowrie were not part of the elaborate ritual of presentation at court and the attendant series of balls and social events that constituted the "marriage market" of the late Victorian and Edwardian period.

Mrs. Hutchison tackled the marriage problem by entertaining officers from the ships based at Rosyth, just across the River Forth. Isobel noted in her diary early in 1905 that Mr. Padwick of the *Caledonia* came to tea, and about the same time the *Scribbler* contained the story "A Summer Cruise to the Mediterranean by a Naval Officer." It was not long before Nita and Victor Padwick were married—Mama had attained her objective. Padwick was a paymaster in the Royal Navy, but members of the family did not consider him Nita's intellectual equal.[18] The Padwicks' life was governed by naval postings, at first around the south coast of England and the Island of Jersey, later in South Africa and China. Much of Nita's life was spent in rented accommodations, alternating between the south of England and the north of Scotland.[19] In light of the realities of Nita's marriage, Mrs. Hutchison ceased to push her other daughters in that direction. She had also realized that Isobel's strong streak of independence could not be bent to her will.

From early childhood Isobel enjoyed long, solitary walks with the family dog along the river or through the various woods on the estate. Walking was freedom and independence, a time for observing nature, for collecting plants, for turning words into poems combining thoughts about spiritual matters with the beauty of nature. In 1904 the family was on a late summer holiday in the Highlands, staying in Kingussie on the River Spey close to the Cairngorm Mountains, when she recorded in her diary: "Went long walk—15 or 16 miles—longest I have been, lovely scenery." Just a few years later, when Isobel was twenty and Hilda seventeen, they made an ambitious hike through the Highlands, packs on their backs, covering the hundred miles from Blairgowrie to Fort Augustus. Their route, through the heart of the Cairngorms, held some of the loneliest and most striking of Scotland's mountain scenery. There were hotels at convenient stopping distances along the route, but on the Larig Pass (Lairig Ghru) they were caught by bad weather and spent the night with a gamekeeper in his bothy. The experience provided material for a story in the *Scribbler*, as well as a Nature Notes column including a description of the plants found in the pass, with their Latin names.

Shortly after this 1909 expedition Hilda, who had shown considerable talent for music at Rothesay House, went to Paris to study at the Sorbonne, living in the Latin Quarter. Music was an acceptable field of study for young ladies, and Mrs. Hutchison approved. Isobel, shy and reserved by nature, appears to have been content to remain at home.

Living at home as an unmarried daughter in a middle-class family has often been a breeding ground for social tension. Florence Nightingale, confined within a Victorian home and not permitted to pursue a nursing career, had felt condemned to a life without purpose. The greater importance attached to the education of a brother was a further cause of tension within Victorian families.[20]

Having proved herself bright and capable at school, Isobel made the best of her situation by taking short occasional courses in Edinburgh, reading widely, and concentrating her efforts on trying to publish some of the poetry that flowed steadily from her pen.

Writing, both prose and poetry, was one of the accepted occupations in the *Dictionary of Employment Open to Women,* published in London in 1898, and at the turn of the century several hundred women in Britain were writing for a living. In the quiet backwater of Carlowrie, Isobel seemed unaffected by the Edwardian ferment that was sweeping away some of the stolid attitudes and rigid conventions of the Victorian period. The shift reflected the change in the monarchy and was more of style than substance. The suffragist movement was the only obvious manifestation of change, and Isobel gave no hint of interest in its activities.[21]

She and Nita made separate visits to the Continent and wrote travel articles and other stories for the *Scribbler.* Hilda, now a serious student, contributed musical compositions or dissertations on French literature to the family magazine. Walter, on the other hand, studying in Edinburgh to qualify as a chartered accountant, was no longer part of the editorial staff. Writing was such an essential part of their lives that Isobel and Nita continued their teenage magazine even though they were now in their twenties.

In what turned out to be the final issue of the *Scribbler,* the 1911 coronation issue, the forthcoming travels of the three Hutchison sisters were described. Isobel and Hilda would be spending the winter and spring in Rome, living with an Italian family to learn the language. Nita, who sent an article about the island of St. Helena and a poem on sailing south, was already in Cape Town on a two-year tour of duty with her husband. In a high-spirited editorial decorated with cartoons, Isobel looked forward to producing an "Italian Special" the following summer. This issue was never produced. The deck had been shuffled, and fate was about to deal the family a hand containing two death cards. Never again would Isobel's writing be quite so lighthearted.

2

The Search for Meaning

Early in her novel, *Original Companions,* Isobel reported a conversation in which her father suggested she might like to be a gardener when she grew up. She replied, " 'I want to be a poet,' voicing a hitherto secret ambition with a violent blush."[1] Later in the novel she again expressed the desire to *be* a poet, although she did not expect ever to *write* real poetry. Already Isobel was sufficiently steeped in great literature to recognize true poetry, but her innate modesty prevented her from considering her own attempts as anything more than inspired verse.

The child who expressed the ambition to be a poet would not have understood the full import of her statement, with its links to the world of magic, myth, and dreams. The novelist recording the child's remark already knew the intensity of the commitment to finding the absolute words to distill one's innermost feelings into a few lines of verse.

Isobel did not merely desire to be a poet, she wanted to be published and earn an income from writing. Having an allowance provided by the interest on her inheritance enabled her to aspire to this difficult and financially unrewarding career. In 1910 she began a "Literary Venture" notebook recording the fate of each poem or article she submitted for publication; her first small success came the following year when seven poems were accepted by various newspapers and journals.[2]

In the twelve years since Thomas Hutchison's death, the family at Carlowrie had recovered from the loss and continued to live in the manner he had foreseen when he made provision for them. Even though the grief remained deep inside Isobel like a cold stone, the passage of time had so hidden it that she was hardly aware of its presence. In the summer of 1912 a calamity struck and revealed the pain still there. Her brother Frank, age sixteen, was climbing in the Cairngorms in August. Trying to reach an inaccessible plant, following the passion of his father and his sister, he fell to his death.

On this terrible event the diarist is silent. A veil is drawn across the scene. The diaries for the years surrounding Frank's death do not exist. The poet, however, continued to distill her private thoughts and feelings into words. Some of the poems were published, including one that was awarded a prize by the *Westminster Gazette* in 1913.[3]

The waves of patriotic fervor that swept the British nation in 1914 touched even the sleepy village of Kirkliston. With the declaration of war on August 5, Isobel began once more to record events and feelings, this time in detail in a journal.[4] Walter was mobilized as a member of the peacetime Territorial Army, and his sisters proudly watched him march out at the head of his company on that first day of the war. Walter's unit was stationed for training along the coast at North Berwick, a beautiful seaside village where the Firth of Forth becomes the North Sea.

For a few months Isobel noted in her journal the effects of war as felt in the area of Edinburgh and Kirkliston. Horses were commandeered for the army, German spies caught near the naval station of Rosyth were imprisoned in Edinburgh Castle, and guns were heard on the Forth—all within the first week. A military funeral for a German prisoner and the report of a ship sunk by a mine at the mouth of the Forth brought the war closer.

Isobel and Hilda reported to the Red Cross office in Edinburgh and were assigned the task of writing letters for the illiterate dependents of soldiers already sent off on active duty. Coming into contact with Edinburgh slum dwellers, Isobel tasted an entirely new slice of life, and she used her powers of observation to describe them as though she were writing a novel. This led her to examine her own life: "I seem not to have a practically useful life. I am a queer mixture of practicality and idealism. Would spend life in contemplation given the choice—but the practical details are in my head when I need them."

Women of Isobel's class had not been educated to lead practically useful lives outside the home, since they were expected to remain attached to their families until they married. Yet by 1915 women were being employed in office jobs to replace men, and by the summer of that year they even worked in industry. Women from domestic service, dressmaking, and low-paying industries flocked to the munitions factories, but few women of the upper and middle classes joined their ranks.[5]

Though surrounded by wartime bustle, Isobel felt she was hardly affected by the prevalent mood of the people around her but was living mentally on a different plane. Seeing for the first time the harsh realities that many people faced daily, she recognized her own privileged and sheltered life and felt a twinge of guilt at being able to use her poetic imagination to float above the turmoil. The suggestion of spending her life in contemplation evokes the image of becoming a nun, but with her strong Presbyterian background this option was not open to her.[6]

By the beginning of November 1914 the grim realities of war had become too painful to let Isobel continue her objective reporting in her journal. After that date the pages are used for poems and ideas for poems.

Wild places continued to be her source of inspiration, although after Frank's death Isobel did not return to the Cairngorms but took holidays on the seacoast of the West Highlands and Western Islands. In the summer following the tragedy, Isobel found peace and serenity on Tiree, the outermost island of the Inner Hebrides, low, green, and treeless, "circled by broad beaches of silvery sand" where the "breakers bear with them an impression of tremendous strength and happiness."[7] Early in September 1915 she and Hilda sailed again to Tiree from Oban after walking there from Doune, a town near the entrance to the Trossachs. This seventy-mile route was easier than their previous hike through the Cairngorms and was well provided with hotels and ferries or rowboats to cross some of the lochs. They reached Oban in six days, climbing to the top of Ben Cruachan on the way.[8]

To his mother's dismay, soon after the war began Walter volunteered to serve abroad if needed. The early course of the war made it obvious that every available man would be sent to the battlefields, but there was a reprieve when a broken ankle prevented Walter from embarking for France with his company and he remained behind at North Berwick. By a strange twist of fate, in November 1915, on night patrol in thick fog, Walter fell from a wall and died of his injuries.[9] The newspapers were filled with the names of sons and brothers of other families dying as heroes in France while Walter, the close companion of Isobel's childhood who had filled the void left by her father's death, was snatched away by a seemingly senseless accident.

Once again the diaries are missing. Isobel turned to poetry in her search for the meaning of life and particularly of death. In 1916 she selected her best poems and published them herself under the title *Lyrics from West Lothian.* A tiny review in a local paper on April 14, 1916, gave qualified praise: "If any reason besides the artistic excellence of the work in this volume is required to commend it to public attention, it is to be found on the title page, that the proceeds go to the Red Cross Fund."

The following year a second book, a play in verse, *How Joy Was Found,* was published by an established press. The Scottish magazine *Field* and the *Glasgow Herald,* both of which regularly published her poetry, praised the childlike lyricism and West Highlands humor and beauty of the work. Not biased in her favor, the *Times Literary Supplement* recognized the ingenuity of her interpretation of the allegory and "the bird-like, spontaneous quality of true song" with "gleams of beauty and truth" in the verse but gave the "queer" play a mixed review.[10]

The play was based on a Celtic fairy tale about how Finn and his band of helpers, each with a special talent, defeated the giant who regularly stole away the hero's children. Isobel treated the tale as an allegory and turned it into a morality play in verse. Finn is humanity, the helpers with their special talents are the virtues of duty, obedience, love, truth, and so on. Although Isobel treats the giant as "a mere notion" without naming him, he represents death, and the giant's dog, which is never seen, is fear. Within the play Isobel gave herself the role of the climber—faith—so accurately self-defining that her voice can be heard in the lines she has given the character to speak. With an awareness of the circumstances of the poet's life, the play can be read as her struggle toward understanding the deaths of her brothers. In spite of the underlying tragedy of loss, the play is a joyous affirmation of her belief in God and the beauty of nature.

Having made a positive beginning as a poet, Isobel embarked on a new course. In 1917 she enrolled at Studley Horticultural College for Women in Warwickshire. Was she following her fictional father's suggestion that she become a gardener? There is more autobiography than fiction in her novel. Isobel's father was nearly fifty years old when she was born and was a quiet, introspective man who loved his gardens and his books. As a young child she made her closest connection with him in the garden and in her own garden plot that he encouraged her to plant. Was she concerned about the lack of practicality in her life and prompted by the patriotic wartime need to grow more food on the Carlowrie estate? Or did she feel the need to study within a structured curriculum? Since completing her Edinburgh schooling, Isobel had never ceased to study in an informal way, whether it was attending lectures in psychology or learning Greek by herself. The Horticultural College offered the opportunity to leave the shelter of home and study a subject that was useful as well as being close to her own interests.

The college was the brainchild of the countess of Warwick, heiress to a fortune and a title in her own right before her marriage to the future fifth earl of Warwick. A woman of great beauty, high spirits, and restless energy, with close ties to the royal family, including an intimate relationship with the Prince of Wales, the countess of Warwick was a leading hostess of Victorian society.

One of Lady Warwick's ideas, Studley College began in 1898 as the Agricultural School for Women, affiliated with the Oxford Dairy Institute at Reading. The purpose was to give educated women training in light farming, gardening, and other rural occupations, to provide the daughters of professional men with the opportunity to earn a livelihood away from cities, as part of a larger movement for women's social reform. The daughter of Queen Victoria, dowager empress of Germany and queen of Prussia, agreed to be the patron of the school, and other powerful friends—Herbert Asquith, Winston Churchill, and Cecil Rhodes—served on the governing council along with peers of the realm, professors, and pioneers for women's rights. On eleven acres of ground, with three houses serving as hostels and with fruit trees, poultry, greenhouses, and demonstration gardens, the school flourished during its first five years.

Deciding in 1903 that more extensive grounds were needed, Lady Warwick bought Studley Castle near the Warwick estate and relocated the school. The grounds, including two farms, had been neglected, and the clay soil was difficult to work. Lady Warwick borrowed heavily to finance the move, but her public appeal for funds generated little support, putting the college on shaky financial grounds for much of its future. At first Lady

Warwick was closely involved in the life at Studley, often being driven over by her chauffeur in an open car and inviting the students to events at Warwick Castle. But as new enthusiasms took hold her interest in the school waned, and her direct involvement became erratic and disruptive.

By 1908 a remarkable woman had become warden: Dr. Lillias Hamilton, whose medical career had taken her to South Africa, India, and Afghanistan. An Australian of magnetic personality, contradictory, controversial, and unconventional in dress and manner, she was a relentless pioneer for women's social reform who held visionary ideals. She leased the property from the Warwick estate in 1911 and formed a board to run the college. Lady Warwick, no longer involved, lost interest in her brainchild.[11]

In 1917, when Isobel arrived at Studley College, it was housed in a sham castle on 340 acres, with a staff of sixteen and fifty-three students. Before the war, the college had attracted students from Poland, Russia, Switzerland, and Japan. The grounds included a formal French garden, herbaceous borders backed by rhododendrons and other flowering shrubs, greenhouses, and even a jam factory where students learned American methods for canning fruit. There were sheep, a dairy herd providing the milk for cheese making, and a two-horse team for plowing. The students were taught marketing, received some business training, and learned enough carpentry to build their own greenhouses.

The war years were difficult everywhere, and at Studley they were marked by shortages. Young women replaced men in the stables, on the farm, and in maintaining the machinery. Horses were requisitioned by the army, and there was a chronic shortage of fodder for all the animals. Dr. Hamilton, who had been away on war service, returned to Studley in 1918 to find things in a terrible state, with everyone depressed and overworked.

During that year the worldwide influenza epidemic swept through the college. In her diary Isobel recorded being assigned to help nurse Doris Bowes Lyons, a young woman from a titled Scottish family, who died the following day.[12] Within a few days Isobel had contracted the virus, and it was feared that pneumonia would develop. But her strong constitution pulled her through, and two weeks later she was able to return to classes.[13]

While she may have been physically strong, Isobel was still emotionally delicate and, like her father, somewhat unapproachable. At the beginning of many of the diaries of those years, Isobel methodically listed what she hoped to accomplish by the end of each year. In 1917 and 1918 some of the goals she set for herself were rigorous but practical, while others show the depths of her insecurity and loneliness. Each list contains about fifteen items, in roughly three categories: mental, physical, and personal improvement. She began by outlining the writing she intended to produce—poetry, drama, prose, and even humor—always with the hope that one volume would be published by the end of the year. Her extended visits to Tiree made her want to learn Gaelic, where it was the common language, and to this she added Hebrew one year and Greek the next. The two classical languages would help in her ongoing study of the Bible. In the physical category, each year she proposed to make another long walking tour, continuing the practice she had begun a few years earlier. With regard to personal development, her resolutions are revealing. Each year she hoped to lose all fear and self-consciousness and planned to try to widen her sympathies all around. She resolved to devote her full attention to whatever she was doing, however tedious, and to enjoy everything thoroughly, seeking delight as an end in itself. One metaphysical resolution recurred each year: "to unite matter and spirit."

Close to the top of her annual list, following the literary aspirations, was the hope that she would make one good friend. Such a disciplined, ambitious person of high-minded introspection, despite the determination to be cheerful and agreeable at all times, would not be open to easy camaraderie or superficial friendships. To become a friend of Isobel Hutchison's, one would need special qualities. By the end of her second year at Studley College she had found such a friend in Medina Lewis.

Medina Lewis, ten years younger than Isobel, grew up in Glan Hafren, a comfortable Georgian farmhouse on a forty-acre estate in Wales. The two women had similar backgrounds. Medina's father was a landowner who ran the family business, sheriff of the county, and active in politics. Her maternal grandfather was a sea captain. The five Lewis children were educated at home by a governess in an idyllic country setting until they were sent to boarding school. The family had suffered the tragic loss of their oldest child, a son, who died at school from an asthma attack. In spite of this, childhood years at Glan Hafren seem bathed in a golden light, whereas even the happy years at Carlowrie are reflected in subdued pastels—the difference between Welsh vivacity and the reserved and dour Scottish temperament.[14]

Medina was someone who could understand Isobel's literary aspirations. Both her parents were well educated, her father a graduate of Cambridge. More remarkable, in 1894 her mother received a master's degree in philosophy with first-class honors from the University of London. British universities were slower than North American ones to allow women full participation; Oxford granted its first degrees to women in 1920. Mrs. Lewis had studied at the University College of Wales but had to take her examinations at the University of London, where she was the only woman in the group of ten candidates.[15]

The Lewis family had a rich social life. In addition to the influential people of the county, their friends included George Meredith, the Victorian novelist, and J. M. Barrie, playwright and author of *Peter Pan*. Medina's sister, Eiluned, was a budding writer, and her charming 1934 novel *Dew on the Grass*, still in print, is a fictional portrait of childhood at Glan Hafren from which it is possible to draw a sketch of Medina.[16] As the older sister in the novel, she is ever helpful, kind, thoughtful, and responsible, always there when needed. Medina did not lack for friends, but for Isobel her friendship was unique and vital. During the years at Studley College, Isobel's self-confidence developed from having a friend outside her close and extended family circle. Although their lives diverged after Studley, with Isobel becoming a world traveler whereas Medina was needed at home for many years as a companion to her widowed mother, it was Medina's meticulous care in sorting Isobel's papers after her death that preserved them for posterity. Unfortunately, no correspondence between the two exists—Medina had too often fallen heir to the task of sorting the papers of others to keep her own letters.[17]

Isobel was twenty-eight years old, a published poet, accustomed to a life of quiet independence with sisters, cousins, and books as her closest companions. At Studley she was suddenly thrust into sharing accommodations, study, and work with young women who were from similar social backgrounds but roughly ten years younger. At first her natural reserve created barriers, but gradually she became part of a group enjoying the normal activities she had missed at their age. Isobel was swept along in a ferment of excitement. Her college friends became her "family," and they visited each others' homes or took holidays together on the coast of Cornwall and the island of Tiree, with Isobel's sister Hilda occasionally joining the group. The object of Isobel's chaste affection was not Medina but an Australian girl whose experience of life was much broader than her own. Normally shy and restrained, discomfited by even the most innocent physical contact, Isobel was thrown into a turmoil of conflicting emotions and confused feelings such as she had never experienced before.[18]

As Martha Vicinus notes, "Boarding-school life was built upon a series of paradoxes: distance and intimacy, self-development and repression, discipline and freedom all combined to encourage a self-sufficient, intensely emotional and achievement-oriented world."[19] Intense relationships were inevitable within the closed world of the single-sex boarding school or college and were sometimes encouraged as a preparation for life beyond the school. However, while on holiday with her college friends, who suddenly found men more entertaining than their own group, Isobel confided to her journal: "I'm afraid I'm not really feminine after all—feel as if I belonged to neither sex—a sort of onlooker."[20]

Despite her personal identity crisis, Isobel attained her certificate in horticulture and promptly enrolled as an occasional student in theology at King's College, University of London.[21] She registered for courses in comparative religion, mental philosophy, and Old and New Testament studies. A bed-sitting-room in a women's hostel on Inverness Terrace, off Bayswater Road, became her temporary home in central London. Her room had a gas ring on which she could boil a kettle, and there was a dining room where she met the other residents. In contrast to her friends at Studley, the residents were independent older women with strong opinions, supporting themselves in creative occupations; they included a singer, a stained-glass artist, a drama critic, and even a surgeon. Once again Isobel had to overcome her natural shyness and insecurities to find her way in new circumstances. The emotional ferment begun at Studley was compounded by the stimulation of her new life at the university and in the hostel.[22]

The period in London, beginning in the autumn of 1919, lasted only half a year and was marked by fevered activity and emotional instability. In addition to reading widely for her courses and attending complex theological lectures where male students predominated, Isobel summoned up courage to visit publishers with her poetry. London offered a wide spectrum of opportunities for broadening her experience, and women from the hostel invited her to attend evening meetings on subjects as varied as Christian Science and spiritualism.

She began work on her novel in the form of letters to her "Heavenly Father," discussing with him all that was going on in her daily life as well as her thoughts and feelings, always probing for meaning. Although she claimed to be short of money and was economical about what she spent on food, she created anxieties for herself by giving away her last shilling each week to needy strangers sought out on the street corners of London.

Even in later life, some of Isobel's most momentous decisions were directed by impulse, so it is not surprising that in her unbalanced state she instantly accepted suggestions. The offer of an assisted passage to go to Australia as a farm girl set her dreaming of travel to distant places, even though her aim was to be a writer, not a farmer. Mrs. Hutchison discouraged that idea and enlisted the family lawyer to persuade her that the Australian climate would not suit her. Likewise, praise from a publisher telling her she should give up study and live as a poet on a Hebridean island prompted an immediate wire to Tiree to rent a cottage. During Isobel's time in London Hilda was a frequent visitor, and their sister Nita briefly rented a flat near her, possibly at the instigation of their mother, alarmed at the strange tendencies Isobel was exhibiting.

While in London, Isobel discovered the play *Mary Rose,* by J. M. Barrie, and went to see it four times over the course of a few weeks, taking any friends and relatives who were in the city. The play is based on the Celtic legend about Kilmeny, the pure maiden who mysteriously disappears, taken by fairies, and returns to recount the visions she has seen, unaware that she has been away.[23] The two ideas of maidenly purity and escape from the earth to visit heaven struck profound chords with Isobel. The play and the legend resonated so deeply that many years later she used them to explain the strong impulse that drew her to northern travel.

During May and June of 1920 Isobel's spiritual fervor reached a climax. In the euphoria resulting from a request for some of her poems for a supplement on Edinburgh poets in the *Poetry Review,* she began to feel the hand of God in every coincidence, directing her every movement. God began speaking to her in dreams, revealing the secrets of life and death. After days in a state of ecstasy she would be left emotionally drained and physically exhausted. Twice during this period she commented in her journal that people knowing her thoughts would think her mad, and she wondered how to convince them she was not.

The journal for 1920, in spite of the fact that some pages have been torn out, gives a clear picture of a young woman whose psyche is balanced on a knife edge. Before the end of June Isobel left London and returned home to Carlowrie to work on her novel. From the beginning of July the pages of the journal are blank for four months. During that time she was immobilized by a deep depression and lay in bed for several weeks, powerless to move, tormented by thoughts, visions, memories, and voices, feeling helpless and alone. In her novel she described the morning when she believed she was blind (hysterical blindness), then gradually felt a presence in the room and looking up saw the word "Love" written in brilliant light. Recovery began from this moment, and by late fall she was well enough to go to her beloved island of Tiree.[24]

Through the winter on Tiree, Isobel could feel her strength returning. Living close to nature in the pastoral setting beside the sea, she found her thoughts occupied with spiritualism and second sight, a phenomenon fully accepted in the Hebrides. Once again she underwent a period of heightened awareness, dreams with messages, the spiritual sensation of "signs" from God in everyday occurrences, even messages from her dead father and brother Walter. As had happened in London, the period was triggered by praise for her poetry, this time an article in the *Scotsman* mentioning her as one of only four poets worthy of inclusion in an anthology of contemporary Scottish poetry. For two weeks she was swept by feelings of being reborn and of being given the answers to all her troubled questions about death. Such spiritual fire could not continue to burn without causing destruction; fortunately, after two weeks of intense joy the flame sank to a glow and Isobel was left with a residue of peace. Fear had been banished forever from her soul, and never again was death a source of sorrow.

Returning to Carlowrie in March 1921, Isobel began to work seriously on her novel. As a birthday treat at the end of May she went back to Tiree for a week and then to the tiny island of Iona off the coast of Mull. A friend rowed her across the narrow strait to Mull, and Isobel, a heavy pack on her back, walked across the island, took a ferry to Oban, and continued her hike to Taynuilt. The three-day walk in the solitude of highland glens was an indication of her restoration to physical health.

The year's auspicious beginning on Tiree continued right through to the following New Year's Eve, which Isobel celebrated in Switzerland, on holiday with the family of a college friend. In September of that same year she made her first visit to Medina Lewis's home in Wales.[25] Medina's father had died unexpectedly in the spring, and Mrs. Lewis found Isobel a perceptive and understanding guest. Isobel was able to talk to Medina's mother, with her background in philosophy and her deep interest in theology, in a way that was not possible with her own mother. Of all Medina's friends, Isobel was the one who most interested Mrs. Lewis, and they formed a close bond.[26]

By the end of the year *Original Companions* was finished and sent off to the publisher. Isobel wrote of it in her diary: "Novel is a queer production though a great deal of it is true. So much so that I don't think I want it printed. If it succeeds I will know there is enough Truth in it to please God." While there is enough invention to justify calling it a novel, the author's life and feelings were transparent to all who knew her. When it was published, Isobel received many letters from her friends, who were pleased to recognize themselves in events and situations in the book. The women she had known in the London hostel were enthusiastic about her portraits of them, and their letters to her give no indication that they were aware of the delicate balance of Isobel's mind at the time.[27] In the novel she had bared her soul and laid to rest the ghosts haunting her since childhood. She had passed through a kind of purgatory, and like steel tempered in the furnace, she had been purged of fears and emerged with new strength, ready for any challenge that presented itself. A new life was opening up, and the path would lead her in surprising directions.

3

The Making of a Traveler

In *Abroad,* his critical book on travel writing, Paul Fussell has drawn a clear distinction between travelers and tourists.[1] In general, tourists follow an arranged itinerary with a predictable route to ensure comfort, familiarity, safety, and the opportunity to visit the most famous sights/sites, usually in the company of others. They will see conventional things in a conventional way and will be shielded from the shock of too much foreign culture. Travelers, on the other hand, are self-directed, making their own arrangements and not afraid to deviate from their proposed plan to take advantage of the unexpected. Their objective is to blend in with the local populace, to travel unobtrusively, and to absorb local culture.

A young Englishman, Thomas Cook, began organizing group excursions as early as 1841. A book salesman and Baptist preacher, he was inspired to charter a special train to take 570 teetotalers on the eleven-mile round trip from Leicester to Loughborough to attend a temperance meeting in the English Midlands. From the success of this unlikely beginning grew the ubiquitous Thomas Cook and Son travel agency, which changed the nature of travel and travelers.[2] The aristocratic Grand Tour of the great sights of Europe and the eccentric expedition to unexplored destinations in Africa and Asia were overtaken by the democratization of travel. Thomas Cook simplified the process by introducing coupons that became accepted at hotels throughout the world. Cook personally tested routes and accommodations in Europe and arranged circular tours, dominating the travel market. In the early years Cook's service included having a local agent meet tourists disembarking from the ship or train and conduct them to the hotel. A traveler in Cook's capable hands should not be exposed to any unpleasant surprises.

Especially popular in the 1920s was the tour of Palestine—the Holy Land—with its overtones of pilgrimage. It could be combined with an excursion to Egypt, where Cook owned fifteen steamers operating as luxury hotels on the Nile. The opening of Tutankhamen's tomb in 1922 gave tourists an additional incentive to flock to Egypt.[3]

For Isobel Hutchison a pilgrimage to the Holy Land was a natural start at fulfilling her dream of seeing the world, expressed several times in her novel. She spent January 1923 studying Arabic and making preparations for this first foreign journey entirely on her own. Leaving from the port of Liverpool at the beginning of February, she sailed on the S.S. *Herefordshire,* captained by an uncle of her friend Medina Lewis, who came to Liverpool to see her off. Though she suffered a touch of seasickness in the Bay of Biscay, her voyage through the Mediterranean was generally calm, and the ship reached Port Said at the head of the Suez Canal two weeks later. She spent the rest of February touring in Egypt, visiting the Pyramids and the Sphinx from Cairo, going by train to Luxor and the Valley of the Kings, and returning down the Nile on one of Thomas Cook's boats. Since she was not part of a tour group, Isobel stayed at a YWCA and traveled with parties formed and led by the ladies in charge of the hostel before going alone by boat and train to Jerusalem.[4]

In her original plan, Isobel had considered remaining in Jerusalem to work as a gardener at the Garden Tomb, taking the place of a woman who had been murdered two years before. Her mother emphatically vetoed this idea.[5] After visiting all the required sights Isobel stayed until Easter, saw the Muslim procession on Good Friday, visited the Wailing Wall of the Jews on Saturday, and found a Scottish service at a Christian church for the Easter service on Sunday. With the Easter celebrations over, it was time to move on.

She went north to Nazareth for a week. The hotel was crowded, and for a few days she had to share a room, until she "found a glorious little room all to myself." Best of all were ten days spent beside the Sea of Galilee, in a village she called Galilee. Bathing in the lake, sketching, and roaming hillsides alight with the wildflowers of April were the happiest part of the whole trip.

Isobel thought about extending her tour to Constantinople, but she put it out of her mind and at the end of April began her homeward journey on a ship from Beirut. Progress around the eastern Mediterranean was leisurely as the ship called at Smyrna, Athens, Malta, and Naples. While walking on the promenade deck, Isobel met the Scottish clergyman Dr. Norman MacLean, who had conducted the service she attended on Easter Sunday. He was in Jerusalem in connection with the building of St. Andrew's Hospice for pilgrims, in memory of the Scots who fell in the war. A dynamic clergyman with a magnetic personality, he recognized in Isobel a troubled soul in need of counsel: "Dr. MacLean is trying hard to teach me what it is like to have a *human* father instead of just imagining a far off Heavenly one, and I felt just as if my book [*Original Companions*] had come true and my own father had actually reappeared on sea, if not on earth. . . . [Dr. MacLean] is trying to teach me how to be more 'human' as he puts it, and to know more about the wonder of my own nature so that I can write better and help get [become] better on earth."[6]

Dr. MacLean continued on the ship to England, but Isobel disembarked at Naples and caught a train to Rome and then to Genoa, Paris, and London, spending a few days in each of those cities. She arrived at Carlowrie in the middle of May, three and a half months from the time she left. The generally favorable reviews of *Original Companions* had reached her in Jerusalem, encouraging her to think that someday she would write a great book. While in Jerusalem she had written articles and poems and posted them to the Scottish Women's Guild, the *Glasgow Herald,* and the *Daily Mail.* On her arrival home her diary recorded the glorious spring weather, her pleasure at being reunited with her dog, Mick, and a laconic "All MSS returned, Alas!"

Although Isobel's travel in Egypt and Palestine was done independently and she was not herded from one site to the next by a dragoman, much of what she experienced would be similar to one of Thomas Cook's tours. The disillusion when biblical scenes are commercialized and overlaid with the accretions of competing religious sects, the crowds, the beggars, the smells, and the foreign food were there whether one was traveling alone or as part of a group.[7] People who set off believing themselves to be pilgrims were for the most part only tourists. Other travelers of the time, including Sir Wilfred Grenfell, the famous medical missionary to the Labrador, reported similar disillusionment.[8]

The following year Isobel had an entirely different travel experience. In 1924 she was invited to accompany a wealthy Edinburgh acquaintance of about her own age on a six-week tour of Spain and Morocco. The arrangements were by Thomas Cook, and the two women visited all the important sights in Spain, Portugal, Majorca, and southern France in addition to the main cities of Morocco. Traveling in comfort and unaccustomed luxury but with a companion who proved uncongenial, Isobel was unusually eloquent in her diary:

> Miss G. has quarrelled with me—I was too Scotch and independent . . . so now I am free! What a feeling of relief—never again shall I go travelling at someone else's charge unless I know that someone *very* intimately. I feel as if I could breathe free air now. . . . I always had an uncertain feeling that as soon as our six weeks were up and Cook's guiding hand removed from our tickets and coupons there would be difficulties and I was right. . . . She is another strange *thrawn* nature with keen jealousy and fieriness, yet a nice side too—if only her wealth had not spoilt her so. Money truly *never* impresses me and I think it is even nicer to travel not deluxe, though I am glad to have had the experience of the other to know this. Certainly much nicer in pleasant company than deluxe in unpleasant.[9]

Isobel resolved never to repeat the experience.

She had learned a valuable lesson: freedom was more important than comfort. She had enough of her own money to be self-sufficient, but not enough to be lavish. Isobel was brought up to be courteous and even-tempered, and it was extremely rare for her to quarrel. Only two such occasions appear in her diaries, both times when her independent spirit and "Scotch" nature brought her into conflict with a woman—never a man—trying to dominate her. In each case her response was a carefully considered cool politeness and withdrawal from the situation. Refusing to allow her life to be controlled by anyone else, and being willing to accept the conditions necessary to follow her own path, marked the style her subsequent travels would take.

Walking was a form of freedom and as natural to Isobel as breathing. Although there was a car at Carlowrie, with the gardener acting as chauffeur, it was not unusual for her to walk the eight miles to Edinburgh along the Turnhouse Road, before that road was closed by the building of Edinburgh airport.[10] By 1924 she had made several walking tours in the Scottish Highlands, and in the autumn after her tour of Spain she set off alone, knapsack on her back, to walk the length of the Outer Hebrides.

The roads across the pastoral, windswept land of the Outer Hebrides are rarely out of sight of the sea on the Islands of Barra, South Uist, Benbecula, and North Uist. Even on the largest and most northerly island of Lewis the sea lochs run deep and fjordlike into the land. Trees are rare, and much of the barren land's beauty comes from the clarity of the light reflected from the sea. On Isobel's October walk, sunshine prevailed and the sea was radiant. The peaks on the islands of Mull and Rhum and the Cullins of Skye floated mysteriously on the western horizon.

Local custom decreed the method of crossing from one island to another. To reach South Uist a ferryman was summoned by telephone on Barra. The island of Benbecula, between South and North Uist, is harnessed to those islands; at low tide the water is shallow enough to be crossed on foot. Isobel crossed the ford from South Uist to Benbecula by horse and trap and walked the four-mile ford to North Uist with a wedding party. The track is marked by stone beacons on the rocks, but every few years the route has to be changed on account of the shifting sands.[11] Away from the marked path there is danger of becoming mired in quicksand. According to Celtic legend, fords are the favorite meeting places of ghosts.

The enchanted landscape creates the atmosphere for stories of fairies who live underground or in gaps in the rock and abduct travelers, releasing them years later with no memory of captivity. Isobel had encountered this idea in the J. M. Barrie play *Mary Rose* during her troubled days in London, and it continued to haunt her. In her novel, someone with second sight, another belief of the islanders, had foretold the death of the heroine's brother on Tiree. Her nature was closely attuned to the Hebridean atmosphere of legend and magic.[12]

When the distance between the scattered hotels was too long for one day's walk, Isobel would find a bed for the night at a fisherman's cottage or a farmhouse. Sometimes the hotels were full and she was directed to a home to ask for shelter. At Lochmaddy in North Uist, a solicitor gave up his bed for her rather than see her left on the road. The room was full of books—

Josephus's history of the Jews and Martin Martin's book on his tour of the Hebrides, published in 1703—and she passed a pleasant evening despite the rats that lived in the wainscoting. The shyness that plagued her in the artificial atmosphere of drawing room society was not a problem when she met local Hebrideans naturally on their home turf.

Harris, the southern part of the island of Lewis with Harris, is the most mountainous part of the Hebrides, and the road over the shoulder of Clisham, the highest point in the Hebrides, was wild and grand. But on the last day's long walk, over monotonous peat land, darkness fell early. The moor turned purple and faded into dimness before Isobel reached the Butt of Lewis, the northern tip of the island, and saw the welcome glow of the lighthouse. From it she looked out across the billowing green Atlantic, and the seed was planted in her mind for her next trip—Iceland.

Soon after returning home from the Hebrides, Isobel submitted an article to *National Geographic* on her 150-mile trek, and in June 1925 she was overjoyed to have it accepted for one of the largest checks she had ever received for her writing—$250, or more than £51. The previous year she had received one almost as large for a serial written for the *Westminster Gazette*, but until then her writing had never earned more than £20 in a year.[13] Having been accepted by a prestigious American magazine, Isobel could begin to seriously think of herself as a writer. It would be thirty years before the article was printed in 1954, after she repeated the journey with a photographer, but the acceptance of her article marked the beginning of a long and fruitful relationship with Gilbert Grosvenor, president and founder of the National Geographical Society.

With her windfall from *National Geographic* in hand, Isobel set off for Iceland as soon as she could make the arrangements. Iceland was not on the list of preferred destinations for tourists making their bookings with Thomas Cook and Son. Some of the great British travel writers of the 1920s had been mired in the trenches during the war, and on returning home to shortages of everything that made life sweet, they went off in search of warmth and sunshine: D. H. Lawrence to Italy, Gerald Brenan to Spain, Robert Graves to Majorca, and Lawrence Durrell to Corfu.[14]

On the other hand, Iceland had been a prime destination for British travelers in the late eighteenth and nineteenth centuries. The botanists Sir Joseph Banks and William Hooker were among the first to visit the island

and publish their impressions. The attraction of the Icelandic sagas caused William Morris, pre-Raphaelite artist, craftsman, and socialist writer, to make two extensive camping treks in Iceland in the 1870s. His travels by pony with three friends and two guides are recorded in his *Icelandic Journals*.[15] Isobel's choice of destination was guided by romantic instinct rather than the writings of earlier travelers.

In July 1925 the incoming Danish steamer *Island* was crowded with Danes seeking to holiday in Britain, but there were few British passengers going aboard to make the return trip to Iceland. Isobel was glad to have a cabin to herself; she took medicine to prevent seasickness and settled in to enjoy the three-day ocean voyage. She was attracted to the sea in all its moods, and a boat trip would be the prelude to most of her adventures.

After spending nearly a month in and out of Reykjavik, Isobel had visited all the recognized tourist sights. On an eight-day pony trek she visited the geyser fields and continued halfway up the wild red slopes of Hekla, where she climbed on foot to the snow-piled crater. Hekla, the most frequently erupting of Iceland's volcanoes, had its last major blowout in 1845–46.[16] To escape the hordes of German tourists who landed at Reykjavik on their way to Spitsbergen, she sailed to the port of Akureyri, visited the herring fishing stations of the north coast, and returned to Reykjavik.

Sated with tourist sights, Isobel longed to make contact with the land and people in the way she preferred—a solitary walking tour. The first travel agent she approached laughed in derision. He could not supply a road map of the island because such a thing did not exist, the few roads being mainly pony tracks. There were too many fast-flowing rivers to cross; a guide and ponies would be essential. The final hurdle was that she did not speak the Icelandic language. She was on the verge of giving up her dream and sailing home from Reykjavik when the French explorer Dr. Jean Charcot arrived from an exploratory trip to eastern Greenland on his ship the *Pourquoi Pas?*

Dr. Charcot had made two extensive expeditions to the Antarctic before 1910 and mapped part of the west coast of that continent. The *Pourquoi Pas?* had been built to his specifications in 1908 and was the most modern and comfortable polar vessel of its time. In the 1920s Charcot began making regular oceanographic voyages in Denmark Strait between Iceland and Greenland.[17] On his arrival in Reykjavik, Dr. Charcot was

Iceland

persuaded to give an evening lecture in the town hall about his most recent trip, and Isobel attended. Hearing him speak and having a chance to meet him encouraged her to pursue her dream.

Independence and frugality were an integral part of Isobel's nature. Another facet was her determination to reach a goal once she made a decision, although the way the decision was reached might be highly subjective. We will see that the mere suggestion of an idea was often enough to start Isobel on a journey of some consequence. In this case the name of Dr. Charcot's ship—*Why Not?*—clinched her decision to walk to Akureyri and catch the boat scheduled to leave from that port at the end of August. The distance was 260 miles, and she had fourteen days to complete her walk.

The guide who had planned Isobel's Hekla tour mapped out a route for her, showing where she would require ponies to ford the rivers and recommending farms or parsonages where she could ask for accommodations. The route avoided the great inland glaciers and roughly followed the coastline, crossing the two large peninsulas northwest of Reykjavik. He suggested she make her first stop at the parsonage at Lagafell, only nine miles from Reykjavik.

It was afternoon on a glorious day when Isobel set off with soaring spirits, wearing a tweed suit, carrying a heavy knapsack and rain gear. She reached Lagafell at four and called at the parsonage, where she spent a pleasant hour with the aid of her phrase book. The day was still fine, she was buoyed up with the pleasure of walking, and the route was soon to leave the post road, so there would be no more automobiles. Conscious of the distance to be traveled in a limited time, the temptation to go on was too strong. She chose to continue to the next destination, Mothruvellir, up a narrow track, across a tableland of stones, and over a pass to the river on the other side.

She was shown the route and the first ford to be crossed and warned that it would be midnight before she reached Mothruvellir. As she continued up the winding track in the northern twilight, a mist descended, obscuring the path. Near the top of the pass she met three riders in the fog who told her she had six miles still to go. Close to midnight, farm buildings loomed in the haze. Suddenly a man appeared, giving the impression he was drunk. Isobel, exhausted by her long hike, felt alarmed and vulnerable, but the sound of a woman's voice nearby allayed her fears. She was welcomed into a farmhouse and soon fell into an exhausted sleep. It was the only occasion in all her travel writing when Isobel expressed fear or anxiety.[18]

A walk of twenty-five miles the next day took Isobel to the church and parsonage at Saurbaer. Again it was late in the evening when she arrived because she had missed the path on a high mountain and gone an hour out of her way. During most of the day she had passed no dwellings and met only one party of riders. At Saurbaer she was received with "immediate hospitality" by the wife of the dean and her son, who spoke English. At breakfast he explained that there were two deep rivers to ford that day and she would need a pony to do it. He would go with her for the first twenty miles and then introduce her to another minister who spoke English, who would guide her across two difficult streams on the following day.

Isobel Hutchison's experience of unquestioning hospitality toward strangers is very much the same as that of William Morris fifty years earlier.

The Morris entourage consisted of at least six riders and about thirty ponies. Their arrival at a "stead" or farm, often that of a clergyman, was nearly always greeted with enthusiasm; beds would be found for them, on the parlor floor if necessary, and sometimes a feast would be provided.[19] The next day the clergyman or farmer would cheerfully guide them over the obscure route across bogs, lava fields, and fords. Morris describes many fords where the water was up to the pony's girth and the current so wild that he fully expected to be swept away.

As she approached her next stop at Nordtunga, Isobel was astonished to find that someone had phoned ahead with news of her walk. A Union Jack was hoisted in her honor as she arrived at the farmhouse, which had been turned into a small hotel overflowing with Icelanders come to greet her. Two rooms were opened, with an enormous table spanning them, spread for a festive banquet. Isobel, flustered and embarrassed, was the guest of honor. A lady who spoke excellent English took charge and helped her answer the barrage of questions. She was rescued from spending the night in what had become a public room by sharing a room in the tiny annex, though sharing any sleeping compartment offended her sense of privacy.

Warm Icelandic hospitality continued throughout her walk, just as W. H. Auden reported ten years later in his guide to tourists.[20] Like Auden, Isobel helped rake hay toward the end of her walk when she had time to spare while the valuable summer crop was being harvested. In addition to enjoying contact with the people, she marveled, even in the rain, at the profusion of wildflowers: thrift, polygonum, bedstraw, and golden saxifrage. William Morris, riding over black sands powdered with tufts of sea pink and bladder campion, likened the scene to a Persian carpet.[21] At the port of Akureyri Isobel took a last look back at the glacier with a splash of rainbow on the snow and sailed for home. The Scottish magazine *Field* carried a story of her walk soon after she returned home, and a major article was published in April 1928 in *National Geographic*.

In the two years since she made her prosaic pilgrimage to Palestine as an independent tourist, but a tourist nonetheless, Isobel Hutchison was becoming a traveler. She had left the beaten path to explore out-of-the-way places, meeting the local people and finding accommodations as she needed them. Walking was her preferred way of journeying, giving her freedom, contact with the land, and time to form her thoughts and impressions into words.

Walking across a corner of Iceland had given her such pleasure that she returned five years later and made a longer walk across northeastern Iceland to Akureyri, stopping on her way at Modrudal, the highest farmhouse in Iceland, where she happened upon a wedding party. This walk was the basis of an article in the *American-Scandinavian Review*.[22] The breathless excitement of the timid but adventurous walker in the *National Geographic* article on Iceland now made way for the seasoned traveler.

In the five years between Isobel's two walks across Iceland, much had changed. Tourists with automobiles had appeared on some of Iceland's roads, and the occasional person was even traveling by airplane. An even greater change had occurred in Isobel herself: shyness had been replaced by self-confidence and experience. In the intervening years she had traveled far and had new adventures. The young woman who had shrunk in humility from the enthusiastic crowd gathered to celebrate her first walk now was pleased to join the wedding party's dancing at Modrudal. Contact with strangers in Iceland, brief and hesitant on her first trip, blossomed into admiration and friendship as she spent several days with a family on the way to the wedding feast. Whereas the first trek was made almost on the spur of the moment and was fraught with imagined perils, the later one followed a planned route on which she knew what to expect. Clear, straightforward writing thus replaced the impetuous style of the first article, which was interspersed with stories from the sagas. On the second journey, courage and independence were evident as she forded streams alone in the "wildest mountain wilderness" she had ever seen or refused a lift in pouring rain, accepting only the transport of her heavy rucksack.

Living on an island, she necessarily began each journey with a sea voyage, the time at sea becoming longer as her choice of destination led her farther and farther north. The pattern was to continue. On the boat returning to Scotland, Isobel met two Danes returning home from long service in Greenland. Their descriptions of the ever-changing beauty of that fabled land filled her with determination to see it for herself. At that time Greenland was a protectorate of Denmark, and entry was granted only to those who could demonstrate sufficient reason, scientists and explorers being given preference.

Isobel now had a new goal and faced a challenge to achieve it. But she had enough determination, stubbornness, and willingness to apply herself single-mindedly to any task to overcome the obstacles in her path. Greenland lay just beyond the horizon.

4

Through the Ice Belt

Greenland, the world's largest island, became a colony of Denmark when the Danish missionary Hans Egede, with his wife, Gertrud Rask, ventured in 1721 to search for the remnants of the Norsemen who had settled there about the beginning of the second millennium. With missionary fervor and using trade to fund his ministry, Egede had the tenacity to found a colony where previous attempts by traders had failed. As Denmark's only colony, it was governed by a paternalistic and benevolent government that maintained a closed-door policy to protect the native people. Not even Danish citizens were allowed to visit Greenland except in the service of the state or unless specially authorized.[1] Although the early British explorers Frobisher, Davis, Hudson, and Baffin touched Greenland and mapped some of the coastline on their search for the Northwest Passage, and though Scottish whalers hunted and bartered with the Eskimo in the nineteenth century, the average British person knew nothing of Greenland and its people.

In the nineteenth century all northern people were called Eskimo, and that is the term Isobel Hutchison used in her writing. Within this text they will be designated as Greenlanders, Eskimo in Alaska, Inuit in Canada, and Aleuts in the Aleutian Islands.

To visit the closed and unknown land of Greenland now became Isobel's goal. Besides official Danish visitors, a few persons claiming scientific qualifications were allowed to fill the remaining places on the few ships sailing from Denmark to Greenland each summer. In 1926 Isobel, having studied and collected plants all her life, made application through the Danish consul in Edinburgh to go as a private botanist. The only ship going that summer, however, was already filled with scientists going to Scoresby Sound, halfway up the east coast. She would have to apply again the following year.

Disciplined and orderly, Isobel wrote and studied every day when she was living at Carlowrie with her mother and Hilda. While waiting for permission to go to Greenland she published poetry, as well as travel and nature articles in newspapers and journals, including one article in Gaelic and one in French. To her language studies in Gaelic and Arabic she now added Danish, working with a young man who had lived in Greenland, and was soon reading books on Greenland in Danish. Although her reading looked northward to places she had visited or hoped to visit, including *The First Crossing of Greenland* by the Norwegian explorer Fridtjof Nansen, it also included *Tramping with a Poet in the Rockies* and *Wild Geese,* books that breathed the freedom of unstructured travel.[2]

In her diaries Isobel was still setting out her aspirations for the year concerning her personal development and the work she hoped to accomplish. In 1926 she was less severe with herself than in earlier diaries, emphasizing patience in her work, generosity of spirit, and learning to be fearless of people and events. That year she had the satisfaction of seeing one of her ambitions totally fulfilled. For many years she had been haunted by the Gaelic legend of Saint Bride, the Christian version of the beneficent Celtic goddess. In free and lyrical verse Isobel wrote a mystery play, *The Calling of Bride,* in which her own metaphysical ideas of the spiritual fourth dimension were woven into the legend with dramatic effect. It was published to critical acclaim from drama and poetry critics alike, the first press run was sold out, and best of all, the play was performed by schoolgirls, just as she had envisioned it.[3]

Affinity for all things Celtic prompted a springtime visit to the Hebrides in 1926 for a month of swimming, sketching, and studying the flora. While there Isobel, always open to suggestion, heard a chance remark about the beauty of Norway's Lofoten Islands. "Preferring the wild places of the earth," she set off in August by way of the Faroe Islands to Bergen, Norway, where she found a boat to the islands, north of the Arctic Circle. The Lofotens, warmed by the currents of the Gulf Stream, are sheltered from the prevailing winds by the Alpine peaks that form what is called the Lofot Wall. The fishing villages on Vestfjord have a temperate climate and are the center of the Norwegian fishing industry. The fjord is the spawning ground for cod, and for six weeks each spring the strait is crowded with boats and the air redolent of fertilizer being made from fish offal. By August this was not a problem, and Isobel explored to the outer limits of the islands, collected sixty species of plants, sketched, wrote verses, and climbed, finding accommodations as she needed them.[4]

At last, in April 1927, after some help from the Danish consul, the long-awaited letter of permission arrived from the Danish Foreign Office.[5] Greenland was one step closer! The spring had been a whirl of activity. Poems and good reviews continued to appear in the Scottish press. On a visit to the Lewis family in Wales, Isobel had given an impromptu talk on Iceland—her first public speech. She had even attended the Clan Robertson ball in Edinburgh and danced every reel. Now the reality of the Greenland trip was upon her, the crowning touch to a glorious spring! Only Mrs. Hutchison was less than enthusiastic. She could not understand why Isobel chose such seemingly inhospitable places, unlike Hilda, who had just returned from Hungary.

Although every Victorian mother hoped to see her daughters suitably married, there are several reasons why Mrs. Hutchison did not persist after engineering the marriage of her oldest daughter. Nita's marriage to a naval paymaster, posted here and there about the world, was not all that Mrs. Hutchison hoped. Then, with the outbreak of the First World War, eligible bachelors disappeared into war service and were killed in such numbers that there was a distinct shortage for a generation. At the same time, the accidental deaths of both her sons robbed her of her spirit. While these are all valid reasons, the overriding factor was the independence with which Isobel quietly but firmly pursued her goals, and they did not include marriage. Mrs. Hutchison could do nothing against such a strong will, couched within a sweet and gentle disposition.

Isobel began to collect clothes for the trip. She ordered a new wool suit to be made with breeches to match, and her mother lent her leather jacket, despite her misgivings. Expecting to be gone about six weeks, Isobel bought a pair of black silk pajamas for the feminine luxury of silk combined with the practical color of black. As she said in her diary, "to last me all the time perhaps." For good measure she added a dark brown pair of coarser material. Her diary has an air of anticipation as she rushed to complete writing in progress, read more about Greenland, visited relatives, and prepared for the journey. If there was apprehension it was neither recorded nor evident. At the last moment she discovered that the boat sailing from Edinburgh to Denmark was leaving a day earlier than expected, and she packed in a flurry of excitement, not forgetting to put in her lidded metal collecting box, or vasculum, and her plant presses.

Hilda and Isobel's friend Medina accompanied her to the docks at Leith, running to the end of the pier to wave until the ship was out of sight.[6] Shortly after reaching Copenhagen on July 15, 1927, Isobel learned that her return boat would not be leaving Greenland until November. Rather than arranging to connect with one that left sooner, she quickly made up her mind to stay the extra three months and assumed that the arrangements in Greenland would fall into place. Flexibility and faith that circumstances would unfold as they should were part of Isobel's travel equipment.

The ten days in Copenhagen sped past as she visited the Botanic Gardens to photograph Greenland plants and obtain a book on Greenland flora, went to the National Museum to see recent archaeological finds, and went to the bank to arrange her money. After paying the Greenland Office out of the £100 sterling she had brought, she had £25—about $100—to last the rest of the trip. The French ship *"Pourquoi Pas?"* which had inspired her Iceland walk, was in port, and she saw Dr. Charcot having tea with his friends.

Also at the Copenhagen docks was the *Gustav Holm,* ready to leave on an expedition to the northeast coast of Greenland. There Isobel met the Danish explorer Ejnar Mikkelsen, who was based in Angmagssalik, working with the Danish government to open up new settlements for the people of eastern Greenland. Mikkelsen had made a heroic trip on the Greenland ice cap in 1909–13, as well as an extensive trek across the sea ice north of Alaska in 1906. It would be many years before they met again, but they would remain in contact by letter for the rest of his life.[7]

At last, near the end of July 1927, the five-hundred-ton, four-masted wooden ship *Gertrud Rask* sailed from Copenhagen, and Isobel was on her way. The ship was making its annual trip with supplies for the tiny Danish settlements, strung like beads around the perimeter of the massive island. There were nine passengers for the five cabins, six men and two other women. Five were Danes or Greenlanders involved in administration of the colony, and the others were scientists—a Romanian, a Belgian, and a Norwegian—going to spend the year studying forestry, botany, geology, meteorology, and archaeology. Isobel shared a cabin with a native Greenlander, Oline, returning home to her husband after completing a nursing course in Denmark. She communicated with her in rudimentary Danish supplemented with many smiles.

Her relationship with Oline, though cordial and friendly, was compromised by having to share a cabin—always a problem for Isobel, who not only cherished privacy but was also uncomfortably modest. Unfortunately Oline suffered from seasickness, "Greenlanders being usually bad travellers," as Isobel observed. She herself had her reliable Mothersills remedy. As Isobel became more fluent in Danish, communication improved, but it was easier to develop a camaraderie with the three scientists, who spoke a mixture of languages and shared her interests.

The ship was heading for Angmagssalik, the only community on the east coast, just below the Arctic Circle. With fair winds, much of the voyage was made under canvas. But ten days out from Copenhagen, as they approached Greenland, they encountered the ice belt, a continuous river of ice that breaks away from the northern glaciers and streams south, sealing off the east coast from the rest of the world for ten months of the year. This ice barrier had effectively isolated this part of Greenland from the Danish colonies on the west coast until late in the nineteenth century, when two Danish expeditions ventured around Cape Farewell by umiak. They reported a population still using stone-tipped harpoons but in serious decline from periodic starvation. Conscious of its responsibility to the Greenland population, Denmark established a settlement at Angmagssalik in 1895, provided the people with more efficient hunting gear, and built a trading post where they could obtain supplies when hunting failed. Improved living standards brought such an increase in population that another settlement was established in 1925 in Scoresby Sound, where Ejnar Mikkelsen was assisting a number of families with the move north.[8]

It was this icebound coast that Isobel was approaching in 1927. The wooden *Gertrud Rask* cautiously pushed and ground her way forward through fog and ice. "Once the veil about us is rent by a sudden roar as of artillery; an iceberg close beside us calves with the thunder of an avalanche. So close is it that although it is scarcely visible we can hear the sea hissing and seething against its shattered sides."[9] The siren began to wail as they neared land, signaling the Greenland kayak men in their sealskin boats to come out and meet the ship. There was a prize for the first man to arrive, and he was brought onto the bridge to pilot the ship into the harbor while the other kayaks surrounded the ship like an escort.

It was a proud moment when Isobel went down the gangplank and planted her stout walking shoes on Greenland rock. The people of Angmagssalik, having so little contact with the outside, held to their old traditions, and the Danes considered them closer to their origins than the natives of western Greenland. Without exception, the women still wore their hair tightly skinned back from the forehead and formed into a high conical topknot—a *qilerte*—toward the back of the head, bound with ribbons. Sealskin being plentiful, all were well dressed in silver gray trousers and brightly colored anoraks decorated with embroidery and beading. Isobel thought it possible that she was the first Scottish woman to set foot there.

Anxious to depart before the ice imprisoned the ship, the captain allowed only four days to unload the year's supplies. In the August sunshine Isobel explored the village, a scattering of a dozen houses set on the rocky slopes of a sheltered cove on the peninsula between two deep fjords. She was too warmly dressed for the unexpected summer weather, although to be ready for any possibility she was carrying a bathing suit in her knapsack.

A little river valley was a botanist's paradise of luxuriant alpine flowers—azalea, *Ranunculus glacialis,* and cassiope. Isobel returned to the ship with her vasculum full of plants but her bathing suit still dry. The insect life was out in full force, and her legs were swollen to twice their normal size with bites. The Greenland solution was *kamiker*—sealskin-top boots reaching well above the knee. She soon obtained a pair in exchange for a piece of brightly colored woolen material and some silk ribbon. "The bright bit of cloth called forth guttural exclamations of admiration from the surrounding company of women, all keenly interested in the little bit of business, and the maker of the kamiker shook hands heartily on receiving the stuff, folded it up beside her baby in the sealskin hood on her back, and hurried off at once, as if anxious to give me no time to repent the bargain."[10]

Returning to the ship each night to sleep, Isobel filled her four days at Angmagssalik with collecting and sketching. On Sunday she attended the church service, understanding not a word of Greenlandic but enjoying the harmonious singing. She recognized the hymn tunes and felt as though she were sitting in a church in the Outer Hebrides, "hearing the wash of surf on the white sands to accompany the long slow voices." The Lutheran pastor, Peter Rosing, in black gown and starched Elizabethan ruff, was a member of a distinguished Danish family with a long connection to Greenland.

In the company of a shipmate, Hans Reynolds, a Norwegian archae-ologist, she got into an umiak, or large skin boat, rowed by four young girls. The boat was carrying a load of raw seal meat, and they headed around the promontory into the fjord. The rough seas outside the shelter of the harbor, the remembered warning of the captain of the *Gertrud Rask* that they must be ready to sail on short notice, and the shouts of the girls' mother soon curtailed the expedition. They landed beside the skin tent that was their home, and Isobel and her companion were invited to enter. Isobel described the scene: "A girl, looking like a pile of rags is lying asleep on the flat wooden bench which forms the family bed. Near her an old woman is sitting sewing beside a lighted lamp of 'tran-oil' [train oil; seal or shark oil]. The heat and smell are overpowering. The old woman beckons me hospitably to a seat on the pile of skins beside her, but I retire baffled by the atmosphere."[11] Isobel was not unique in her reaction. On his first visit to an eastern Greenland tent Nansen described the atmosphere as unimaginably unpleasant. He was able to adjust to it, but some members of his party preferred to remain outside.[12]

> An endeavour to trade with one of the older women for a new pair of sealskin trousers proves fruitless, as my Greenlandic is limited to the one word "Ajungilak" ["All right!"]. She thinks I want to purchase the pair she is wearing, and is proceeding very hospitably to come out of them for me, when I interpose hastily. Not even the ethnographical considerations urged on me by my companion . . . would induce me to purchase this poor old Eskimo woman's grey sealskin "bukser" with their fine new scarlet leather embroidery—red-hot, so to speak—from the wearer.[13]

Isobel was covering her confusion with humor. Not only was she too fastidi-ous to accept the sealskin pants direct from the wearer, she would have been horribly embarrassed by the woman's undressing in front of her and her male companion. Her sense of delicacy was not ready to accept the mores of a different culture.

All too soon the only ship of the year was ready to leave. As the residents who had been entertained on board went down the gangplank, there were shouts of "Farewell" and "Merry Christmas." The ship slipped out into the fjord and entered the ice field as sunset painted the glacier in pastel shades of indescribable beauty. They escaped the steady drift of ice from the north into a clear channel and headed around the southern tip of Greenland to Julianehaab and a different world.

Known as the Garden of Greenland, Julianehaab had a farm that specialized in breeding sheep introduced from Denmark in 1916, when the seal population of the area had been decimated by overhunting. Although it was one of the principal settlements of the more populous west coast, in 1927 there were no hotels or hostels. In giving permission to visit Greenland, the Danish government clearly disclaimed responsibility for visitors from their arrival in Greenland. The managers in each settlement had guest accommodations to make available at their own discretion.

Isobel lodged in a guest room at the house of the district manager in Julianehaab and had meals with a bride and groom who were fellow travelers on the *Gertrud Rask*. On the first Sunday evening there was a dance, the Greenlanders stepping out in the best Highland tradition to the music and steps introduced by the Scottish whalers of the previous century. Isobel, an enthusiastic dancer who had attended dancing school in Edinburgh for many years, was captivated by the scene. At first she watched, enjoying the picture created by candlelight reflecting from the brilliant beaded collars of the women with their "Mongolian" faces. But when the music struck up a familiar schottische she could resist no longer and joined the dancers. Thirty-eight years of formally correct life at Carlowrie had made Isobel an aloof observer; it would take time and effort for her to break down the barriers and become a full participant.

The two weeks in close quarters on the *Gertrud Rask* had given the small group on board time to know their fellow travelers. As the only unattached woman of the group, visiting Greenland independently, unsupported by any scientific establishment, Isobel was no doubt an object of some curiosity. She had proved herself to be a good sailor as well as interested in everything and pleasant to all around her. Had she been otherwise, she might have been left in Julianehaab to pursue her botanical quest until the return boat came.

The Norwegian archaeologist Hans Reynolds had come to study the area around Julianehaab, where many relics remained from the Norse settlement of the tenth century. He invited Isobel to accompany him to Igaliko to visit the ruins of the cathedral of Saint Nicholas, built about 1126, the largest ruin in Greenland—now reduced to foundations of sandstone.

After traveling six hours by motorboat up the fjord, they arrived cold and hungry. The Greenlanders were curious about Isobel and crowded around to touch her gazelle-skin coat, lent by her mother, as she followed them up to the house of the *cateket*—the native lay priest. His wife set the table for supper and waited for Isobel to produce the food she had brought. This consisted of a few Oxo cubes, butter, and black bread, fortunately augmented by Reynolds's supply of hard biscuit. Isobel had intended to buy eggs and fish from the natives, but with her Greenlandic limited to one word she had no way of communicating. On the surface it seems surprising that the cateket and his wife did not offer them food. But in this tiny, out-of-the-way community the only non-native visitors would be Danish men on official business who would bring their own food, a subtle form of racism. The couple, awed by entertaining foreign visitors, including possibly the first European woman they had seen, would assume that the food Isobel and Reynolds produced was what they preferred. After a meager supper Isobel went back to sleep in the motorboat while Reynolds put up his tent beside the cathedral.

The following day Isobel explored a field of barley beside the ruins and noted about fifty species of plants introduced by the Norsemen eight hundred years before. The greatest challenge on this expedition came when she was presented with two small salmon trout. Fortunately the young fisherman cleaned the fish. Knowing nothing about cooking, Isobel was forced to improvise with a frying pan and some butter while an interested gathering of Greenland men watched with concentrated attention.

After two days Isobel returned to Julianehaab, leaving Reynolds to his archaeological pursuits. There Isobel was invited to join Dr. Bentzer, a fellow passenger from the *Gertrud Rask,* and the district manager on their annual inspection tour of the tiny outstations on the way to the southern

extremity, Cape Farewell. The area, containing some of the grandest cliff
scenery in the world, had been visited by no other Britons and by few Danes
because explorers of Greenland tended to concentrate on northern regions.
She would have opportunities to botanize at each stop. The motorboat was
run by two young men, nephews of the doctor's Greenlandic nurse, who
not only dispensed the medicines but took charge of the galley as well. The
largest and finest motorboat at Julianehaab had a good-sized cabin with
two padded benches where the doctor and manager slept. Isobel would find
accommodations on shore as a guest, or when necessary she could use the
doctor's tent.

The doctor made a point of stopping at ancient Norse sites along the
way to show Isobel graves and deserted ruins that dated back to the time
of Erik the Red in 985. Familiar with many of the famous names and the
history, Isobel was an observant and interested passenger. She not only had
studied the sagas in preparation for the trip but had gone to the National
Museum in Copenhagen especially to see the most recent relics excavated in
1921 by Dr. Poul Norlund.

With the eye of an artist, Isobel was spellbound by the majestic scenery.
To reach their next stop, Augpilagtok, they passed through silent inner
fjords where the sheer cliffs soared thousands of feet from the water.

> Though almost void of vegetation, save for a splash of vivid moss or
> crowberry lying where some ledge had caught a trickle of moisture, the
> crags were of an amazing variety of colour—black, orange, slate and crimson.
> About their sides the only visible sign of life was the spectral passage of a
> snow-white gull, which floated for a moment against the dark walls close to
> the water's edge, then vanished in the yawning depths of a rock cavern.[14]

At Augpilagtok, Isobel had her first experience of sleeping in a tent,
using a sheepskin sleeping bag on a camp bed. The thrill of bedding down
under the northern lights with the planet Jupiter sparkling above filled her
thoughts with poetry. She woke to the dawn at four o'clock feeling she was
alone at the creation of the world, watching the Spirit of God moving upon
the face of the waters.

At the trading post, Isobel observed the Greenlanders' passion for tobacco when her gift of threepence worth of snuff to a penniless old woman
brought joy and gratitude. Isobel was trying hard to come to terms with the
contrasts she was seeing: the grandeur of the scenery and the utter poverty

of the native people. In her head were the heroic stories of the sagas; at the trading post was life at its most basic.

She was traveling with only a knapsack, but the raggedness of a woman at the tiny hamlet of Nuk moved her to donate her only spare pair of stockings. In contrast to eastern Greenland, here the people were poorly clothed, the seals of the south coast having been harvested almost to extinction. Glad to be able to leave money in the village, she bought a model kayak with a cleverly carved paddler, but not a scrap of material could be spared to clothe the figure.[15]

At Eggers Island, the southern limit of the inspection trip, the seas were too rough to continue on to Cape Farewell, and they returned the way they had come. Isobel was left by prearrangement at Nanortalik as a guest of the Mathiesens, the Danish manager and his wife she had met on the way south. The tiny community of Danes and native Greenlanders lived on a stony island that rose on the seaward side to a summit where the sheer cliff plunged sixteen hundred feet to the icy waters of Davis Strait. In addition to the Mathiesen family there was a pastor, assisted by a native cateket, a *jordmoder* or Danish-trained Greenlandic nurse in charge of the two-story wooden hospital, and a visiting Danish judge making his annual circuit visit from Godthaab. The length of her stay was indefinite, depending on when the ship from Denmark arrived at Julianehaab.

On the first evening the Mathiesens helped Isobel arrange an excursion to visit an inland grove of birch trees, the tallest of the few trees in Greenland, whose fame had brought her to Nanortalik. For the journey she would need an umiak with six rowers and a steersman, as well as a kayak and kayak man, indispensable for safety on any Greenland trip. The excursion would take about five days, and the cost was reckoned at a pound a day. "For little more than a five-pound note, I purchase one of the loveliest weeks of my life."

The merry group set off on a brilliant Sunday morning in early September, happy to be going on a picnic expedition. The Greenlanders of the group consisted of five men, two women—one of whom was the cook and designated "lady's maid"—and one child. The youngest Mathiesen daughter acted as Danish-Greenlandic interpreter for Isobel and shared her tent. The waters of the fjord were clear as glass, mirroring the crystal snow-splashed mountains. At the far end of the fjord the "Isblink" or glacier could be seen spilling down between the dark rocks.

They reached the river leading inland by midafternoon, but the biting insects made them keep moving. The umiak was tracked up the swift, rocky river with the passengers in the boat. Just before they lost the sunlight they reached the shores of the lake where the birch trees were to be found and pitched their tents on a sandy beach. Salmon caught on the way were cooked for dinner. The northern lights or "merry men" came into full play in the starlight, and Isobel drifted off to sleep to the sound of the Greenlanders singing psalms. She felt safe and comfortable with the handpicked group of Greenlanders, yet at the beginning her natural reserve and lack of their language kept her apart. Friendship for her was never instantaneous but was something that unfolded naturally with time.

The morning dawned cloudless and still, becoming almost unbearably hot when the sun rose. As the group went down the lake to find the narrow valley where the birch trees grew, the flies continued to be troublesome. The trees were about twenty feet tall, and the eight-year-old in the party, who had never seen trees, gazed in wonder before trying to climb them. By next day they had returned to the fjord and visited the site of an Augustinian monastery founded by the Norsemen, near the glacier at the end of the fjord. Dr. Norlund had recently excavated in the area and discovered the hewn stone foundations of an extensive building.

After three days in the water the skin of the umiak became porous, and the men turned the craft over on the beach to dry. This was the signal for a rest day, with games, dancing, and singing, the kind of activity Isobel was happy to join. She introduced the Greenlanders to tug-of-war, which they enjoyed with enthusiasm. The return down the fjord took two leisurely days as they stopped to inspect other Norse ruins along the way. On the last night in camp Isobel, now thoroughly at home, visited the Greenlanders in their tent for the first time, and they welcomed her with smiles as she joined in their nightly singing.

The final day held another memorable moment. Soon after they broke camp and set off, a motorboat appeared carrying Dr. Knud Rasmussen, who was visiting Nanortalik for a day on his way to Scotland to be honored with a degree from the University of St. Andrews. Back in the village that evening, Isobel and the Mathiesens dined with Rasmussen on his boat and joined the throng who crowded the schoolhouse to hear him talk about his recent journey across Arctic America.

As he has been described, "Knud Rasmussen was almost a god to all the Eskimos from Alaska to Angmagssalik. His great charm brought him affection from everyone with whom he met. It was only to him that the Eskimo would tell their inmost secrets and darkest superstitions."[16] Born in Greenland in 1879 to a Danish missionary whose wife was proud of her portion of Eskimo blood, Rasmussen spoke the Eskimo language as his native tongue, was driving his own dog team by age eight, and owned his first rifle when he was ten. He was naturally endowed to carry out the four Thule expeditions, named for the northernmost community in western Greenland. Rasmussen, along with Peter Freuchen, had established a base at Thule for exploration and as a trading post to help the northern people. The explorations of northern Greenland had made him a legend throughout the country, and the fifth Thule expedition of 1921–24, on which he visited every native settlement along the coastline of North America, enhanced that legend.

In the remaining two months in Nanortalik Isobel explored the plant life of the rocky island on which the village was built and saw the summer vegetation change to that of autumn. In the company of Mrs. Mathiesen, she observed the workings of a Greenlandic community under the benevolent administration of the Danes. With her she visited Regina, an elderly Greenlander, in her turf hut, taking along hot tea and buttered bread. Regina came to the Mathiesen house each morning to do odd jobs to earn her cup of coffee at the kitchen table. Isobel paid her to sit as a model for a first attempt at portrait drawing. The Mathiesens' relationship to the village was not unlike the noblesse oblige of the British manor house toward the tenants on the estate, and Isobel accepted it without question.

The Mathiesen family was returning to Denmark at the same time as Isobel, and at the end of October they all went to Julianehaab to await the arrival of the ship. Early in November the handsome new diesel-powered *Disko* arrived in port, bringing Christmas trees and gifts for the colonists and loading the cod and salmon that the colony exported to Spain. The ship headed north up the west coast, stopping for several days at Godthaab and finally Sukkertoppen, before turning south for Denmark.

Shipping on the west coast is not subject to the same Arctic current that produces the ice belt imprisoning the east coast of Greenland for most of the year. But during their trip to Sukkertoppen, winter arrived. All the water in the tanks on the ship froze, and no baths were possible. Isobel went ashore at Sukkertoppen to do some sketching and exchanged paintings with a local artist as well as visiting the hospital and sanatorium. On her final evening in port she felt an ominous chill like a cold developing, and the next day she was sick with chicken pox, confined to her cabin and isolated from the other passengers for ten days.

They rounded Cape Farewell on the first day of December, homeward bound in calm seas, only to meet a hurricane on the open ocean. For three days the ship was battered and driven north off course in the worst storm the captain had ever experienced. By December 10 the weather had calmed, and Isobel was well enough to sit in a sheltered corner of the deck, seeing the remote Scottish island of Foula pass in the mists. On Christmas morning the ship she caught in Copenhagen arrived at the Scottish port of Leith to be met by Hilda and the chauffeur, and Isobel was whisked away to Carlowrie.

For Isobel the journey had been magical: "Like Kilmeny, I had returned from Fairyland. Like Kilmeny, I resolved to return thither once more!"[17] Fairyland is frequently mentioned in the first section of her book on Greenland. She refers not to the English version peopled by lightly clad, delicate creatures with gossamer wings but to the mysterious, closed world of the Celtic "little people" who live underground among the rocks and bogs, sometimes kidnapping mortals, who seldom escape. Greenland, the country closed to outsiders, had become her fairyland, and the native Greenlanders, small of stature and living almost underground in stone and turf huts, were her Celtic fairies.

For five months Isobel had traveled in southern Greenland, going where opportunity took her, dependent on shipping schedules and the kindness of others. She was interested in all she saw, was open to suggestions, and so visited places seen by few others, especially on her travels south of Julianehaab, where her imagination and her knowledge of the sagas peopled the ancient ruins. The Danish managers were generous with hospitality, and her presence was a welcome diversion for their wives, who seldom saw another European woman during their stint in the remote fjords. Isobel's good manners and self-sufficiency made her an easy guest, not dependent on others for entertainment. She enjoyed observing the activities around a settlement and was equally happy taking solitary walks to sketch, write poetry, and collect the plants for her private collection at home.

Although she had no connection to the colonial relationship of the Danes to the Greenlanders, she nevertheless wrote about her first contacts with the native people in the superior tone of the outsider. Oline, her cabin mate on the journey out from Copenhagen, "glitters with artificial jewellery—so dear to the simple-hearted Eskimo!" Their hairstyle gives the women at Angmagssalik a "ludicrous air." The village of Nuk is "a miserable collection of turf huts" with a "population as miserably clad as any I saw in Greenland . . . though the people were smiling and cheerful." As contact grew and Isobel learned a few basic words of Greenlandic, she began to see the native people as individuals rather than as an amorphous foreign mass. Her camping trip with Greenlanders to view the birch grove was the highlight of the whole experience, even though on that excursion her upper-class upbringing caused her to designate one of the women "lady's maid" and kept her somewhat aloof from the group in the beginning. On the day of rest and games and when she visited the Greenlanders in their tent on the final night, the barriers were coming down, and Isobel was less an observer than a participant.

At Carlowrie, Isobel's return from the shores of the unknown was greeted with relief. The only communication Mrs. Hutchison and Hilda had received since early in August, when Isobel arrived at Julianehaab, was when she reached Copenhagen in mid-December to wait for a boat to Scotland. Although her mother did not approve of Isobel's outlandish Arctic travels, she could do nothing to stop them. Isobel was now nearly forty, had a small inheritance, and was earning a little from her writing.

Isobel had explored only a minute corner of southern Greenland, but the returning ship had taken her north along the coast. She had seen the austere grandeur of the country, and the North was drawing her back like a magnet. Her feelings flowed forth in a heartfelt poem, "Call of the North":

> *I must go North again! My heart*
> *Is where the white mist lies*
> *About the roots of starlit crags*
> *Beneath the Arctic skies,*

Where through the dusk the Dancers play
Across the northern pole;
I must go North again, for they
Have stolen away my soul.

. . . .

His motherland is there! I feel
Across this southern day
The fingers of the Dancers steal
That plucked my heart away,
They twitter on its strings and cry
All night with starry breath,
I must go North again, for I
Have given my heart to Death, –

The death of snow-white peaks that soar,
Their song hushed on their lips,
So near to very God that o'er
Their rim His radiance slips
And sheets the frosted stars with light
That falls like silver rain, –
His music calls me through the night,
I must go North again.[18]

5

A Householder in Greenland

To Isobel, arriving home on Christmas morning 1927 after a stormy crossing from Denmark, her sojourn in Greenland seemed like a dream. After
unpacking her curios to show her mother and Hilda, she immediately
tackled the task of writing letters and sending parcels to all who had shown
her hospitality and kindness. That done, having come home penniless, she
threw herself into writing articles for newspapers. Before the year was out
she noted in her diary that it was as though she had never been away.[1]
Within a week of coming home she had returned to the same conventional,
narrowly-defined life she had briefly escaped, and she could feel her spirit,
after soaring in Greenland, being brought back to earth and tethered. Life
at Carlowrie continued in the unchanging mold that had been set before she
was born. Isobel had seen another world, tasted new freedoms, experienced
the unknown.

    Shortly after returning home Isobel received recognition as both a poet
and a traveler. She was asked by the British Broadcasting Corporation to
choose nine short lyrics to read on the air as one in a series of readings
by living Scottish poets. Because there was no radio at Carlowrie, Isobel
visited the home of the local minister to hear a broadcast and get advice
on reading in public. The fifteen-minute program took place in the early
evening of March 3. Before that ordeal, she was surprised and gratified to
hear one of her poems read at a meeting of the Scots Vernacular Society, of
which she was an active member. "I sat silently swelling with a face as red
as fire!" Her own article about her Greenland travels, with photographs,
appeared in the *Scotsman* early in January, but she thought it a great joke

to be interviewed at the end of February by a journalist writing an article for the racier *Sunday Post*. She wrote constantly, hoping for publication, yet any form of public recognition surprised and confused her. Hearing from friends who had both heard her broadcast and seen the interview in the *Post* was recompense for the unwonted publicity.[2]

In mid-April Isobel received a reply to her request for permission to return to Greenland. The Office of Greenland Administration granted her permission to visit northern and western Greenland to continue the studies begun in the east and south the previous year and offered her the opportunity to stay through the winter provided she could arrange to live with one of the Danish families.[3] In the same post came a letter from Dr. Rasmussen in London on his way to visit Edinburgh. Isobel immediately enlisted his help in suggesting a family she might live with. Her brief meeting with him over dinner in Nanortalik six months earlier resulted in a firm friendship that was ended only by his untimely death in 1933. Rasmussen, his wife, and his daughter were guests at Carlowrie over the several days they were in Edinburgh.[4]

Isobel had a personality that caught the attention of great explorers. Rasmussen was one of the first of these, but as her horizons widened she could count among her friends many who had traversed the icy wastes of North or South. Though she was modest about her own accomplishments, her intense interest in their work was accompanied by appreciation and intelligence, born out of her own experience and informed by wide reading on the subject. It was rare for explorers to encounter a woman who had firsthand knowledge of the landscape, the people, and the conditions of life in the North. She projected an immediate empathy that fully engaged their attention.

The next few months were crammed with activity. In addition to writing many articles on Greenland, Isobel, who did all her own typing, produced the manuscript of a book on the country, which was ultimately rejected by two publishers. She informed the Royal Horticultural Society of her plans to return to Greenland, and six members contributed £10 each in exchange for seeds to be gathered there. In preparation for the year in Greenland, she was inoculated against typhoid and had an operation to correct a "troublesome" condition of her nose. In addition to buying new clothes at bargain prices, she also bought a cine camera, projector, and film and practiced using them. Even politics engaged her attention, as she worked on behalf of the Conservative Party for the spring election.

Knowing she would be away for at least a year and out of communication for many months, Isobel spent as much time as possible with family members, both at Carlowrie and on short holidays in the Highlands. One poignant day was spent helping her mother go through drawers and choose which of Walter's clothes could be given away. It was thirteen years since Walter had fallen to his death, and his bedroom in the tower remained as he left it. Isobel had come to terms with death at the time of her breakdown, but the dark cloud of mourning still haunted Carlowrie.

Early in July 1928, word arrived from Rasmussen that he had arranged for Isobel to live with the Danish woman doctor at Umanak, the principal community of northern Greenland, above the Arctic Circle. From that moment preparations for departure accelerated. Isobel sailed from Copenhagen late in August on the steamer *Disko,* going directly to the west coast of Greenland. Before reaching Umanak, the ship called at Godhavn on Disko Island, where she met Dr. Morten Porsild, director of the Arctic Research Station. She was thrilled to have the opportunity to visit this important scientific laboratory that Dr. Porsild had established in 1906. He showed her his herbarium and some of his seed collection and discussed the types of plants she would find on Umanak, which he considered a good area for her own collecting.[5]

In mid-September the ship sailed into Umanak fjord and docked at Umanak on the island of the same name. Unlike the long, narrow fjords of south Greenland, Umanak fjord is a wide body of water between two great peninsulas, Nugssuaq to the south and Svartenhuk to the north. A large island, Ubekendt Ejland—Unknown Island—stands like a sentinel between the two headlands. Within Umanak fjord are many rocky islands of varying sizes, of which Umanak Island is one of the smallest. Smaller fjords pass between islands and lesser peninsulas, all leading to the great inland glacier that dominates the scene. All of Isobel's plant collecting during her year in northern Greenland would be within the vast Umanak fjord.

She watched eagerly from the ship as it arrived at 10:00 A.M. and saw a neat, substantial village with well-built storehouses, shops, and houses. Most of the homes were occupied by the ten or twelve Danes who administered the district, although a few of the Greenlanders in the employ

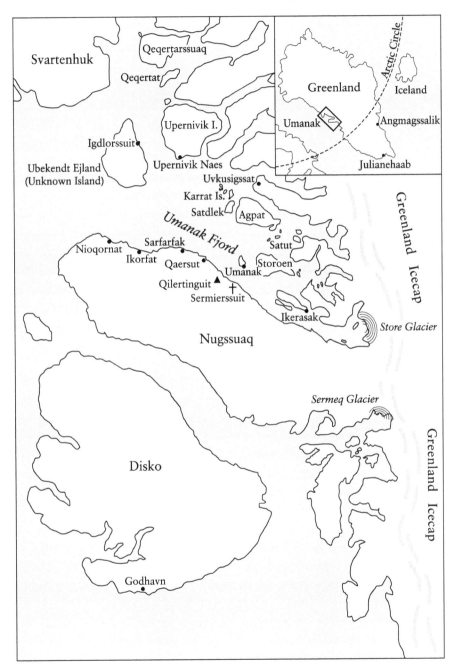

Umanak fjord

of the government had built wooden houses as well. The native dwellings, or *igdlos,* were less obvious. Built low to the ground, flat-topped and made of stone and turf, they held a population of over two hundred.[6] Isobel saw "The colony . . . huddled on the side of an island which at first sight appears to be nothing but an immense rock culminating in a curiously tilted twin-peaked mountain as high as Ben Nevis. The mountain is of reddish colour. This, and its strange shape (like the heart torn from a freshly killed animal) have given it the name 'Umanak'—the heart-shaped mountain."[7] To an eagle soaring overhead, the island would appear as a mere speck of rock in the sea of Umanak fjord, around which towering cliffs rise on all sides to meet the inland ice of the great Greenland glacier.[8]

Isobel was about to take up residence in a paternalistic Danish colony where consideration of what was best for the native population weighed more heavily than the demands of European trade in fish and furs. From the eighteenth century, Denmark held a monopoly on the trade with Greenland, and the regulations controlling access to the colony were imbued with the spirit of commercialism. In the twentieth century the control was less concerned with trade than with protecting the population against the dangers from unchecked communication with the outside world and reserving the basic resources for the inhabitants. By the time of Isobel's visit, some people were expressing concern that the extremely conservative Danish rule was slowing the development of the Greenlanders and leaving them unprepared for a time of free trade in the future.[9]

When she left Scotland Isobel expected to live with Dr. Gudrun Christensen, whose eight-room house was the largest in Umanak. This had been arranged privately for her from Copenhagen by Rasmussen, but on arrival at Umanak the director of the Danish Greenland Office discovered that a three-room house would be standing empty and gave her permission to use it. It was usually occupied by the *baadsmand,* or captain of Umanak's little schooner, who was returning to Copenhagen for the winter. Since all houses belonged to the Danish Office, the house was rent-free, and Isobel had only to pay a small sum to the sailor for the use of his furnishings.

Only a year old, the house still smelled of fresh paint. It perched on the rocks high above the colony and was approached from the front by a steep path from which wooden stairs led to the veranda and the front door. The violent winter winds made this door impractical, and Isobel was instructed to use only the back door, which also opened onto rocks and another perilous path down to the village. The house consisted of a minute kitchen, a tiny bedroom, and a sitting room with a view of dawns and sunsets across the Nugssuaq peninsula to the south.

Isobel was exhilarated by the novelty of being in possession of her own house, which she jokingly named "Scotland Yard." Apart from the brief time away at college, her home had always been Carlowrie, presided over by her mother with Victorian formality. This would be her first opportunity to create a home shaped by her own personality. Since she had no experience of managing a household, however, it was fortunate that every Danish home came equipped with a *kivfak,* or Greenlandic servant, who would take charge of the housekeeping details. Dorthe was the kivfak who came with the house.

Dorthe, a widow with two young daughters, had

> a broad comely face with prominent cheek-bones, kindly grey eyes, black hair parted in the centre and braided behind in plaits—at present insecurely fastened by a couple of my own bronze hairpins. She wears a brightly coloured "anoraq" or jumper, sealskin trousers ornamented down the front with leather piping of rainbow colours, and long scarlet leather "kamiker" or sealskin topboots, coming well up over the knee and finished with embroidered linen tops. . . . I consider Dorthe as ornamental a parlour-maid—perhaps I should say maid-of-all-work—as I shall ever possess.[10]

Proud of being in charge of her own domain for the first time in her life, the words Isobel used to describe her kivfak—"parlour-maid" and "possess"—have an autocratic ring. It appears that she has translated the comfortable, well-regulated life of Carlowrie directly to the tiny house perched on the rocks of north Greenland. But this is a subtle joke, for by the time she wrote her first impression of Dorthe, she knew that "her" kivfak would rule the household and, furthermore, that they would become great friends.

For fourteen shillings a month, Dorthe lit the stove, scrubbed the three wooden floors every morning, made coffee and brought it to Isobel in bed, cooked whale steak, ptarmigan, seal chops, or whatever was on the menu, washed and ironed, baked bread, and sometimes sewed. She also heated water for the daily morning bath in the rubber tub Isobel had brought with her, the water having been melted from chips of iceberg carried up from the harbor by the ice boy. At Carlowrie the staff consisted of housekeeper, cook, parlormaid, and kitchen helpers, their wages paid by the factor who administered the estate. While Isobel may not have known exactly what they were paid, she was aware that the pay she was instructed to give Dorthe was a mere fraction of the amount any one of them would have received.[11]

Expecting to live with the doctor, Isobel had brought mainly luxuries such as chocolate, Edinburgh rock, sweets, shortbread, oatcakes, tins of fruit, and pots of jam. Had she known she would occupy a house of her own, she would have included more substantial fare. She had, however, made provision for Christmas dinner with a tinned haggis and several plum puddings with French brandy to set them alight!

Language was the first hurdle to overcome. Although her Danish was becoming fluent, when she arrived in Umanak Isobel still had only one word of Greenlandic, "Ajungilak."—"All right." When Dorthe had a problem communicating, she would go to the pastor, who wrote Dorthe's message in Danish. The pastor, Otto Rosing, was the brother of Peter Rosing, the clergyman Isobel had met at Angmagssalik the previous summer. He was married to a native woman and was pleased to exchange Greenlandic lessons for lessons in English that came with a Scottish burr. An artist and scholar and a gifted carver of ivory, cultivated and charming, Otto Rosing was a natural leader in the community. Isobel knew immediately that he would be a great friend.[12] His wife spoke no Danish, so their friendship would grow as Isobel mastered the new language.

The second mate on the vessel that brought Isobel to Umanak was an expert botanist who kept a mounted collection of Greenland flowers in the locker under his ship's bunk, collected during hurried visits ashore while his ship was loading or unloading in different harbors. On her second morning in Umanak, stormy and wet, he directed her to the churchyard behind the colony on a productive hunt for plants. She was able to fill a large wooden box with live specimens, to be cared for by the second mate during his return trip to Copenhagen and then dispatched to Britain.

Within the first week the pastor organized a picnic, to give Isobel a chance to botanize and to make the most of the fine weather before the sun disappeared. A Greenland picnic off the island included a night ashore with tents. The destination was the mainland peninsula of Nugssuaq, five miles or two hours' rowing in the umiak, with a crew of seven, a kivfak to do the cooking, and a kayak for safety. Isobel made the most of the opportunity to hunt for plants and collect seeds. The men caught cod and sea scorpions and played checkers as well as practical jokes on each other.

Rain came on during the night, and while they waited for the weather to clear Isobel sketched and the men whittled, all of them paying a visit to the home of a whaler, the only family on that shore. As they were about to embark for the return across the fjord, the beach was flooded by a huge wave. An iceberg had calved with a roar like a cataract, and for five minutes the beach was awash. Had this occurred when they were in the umiak it could have been fatal. Close to that area, at Qilakitsoq, nearly fifty years later hunters found the ancient burial site of eight perfectly preserved bodies—six women and two children, thought to have drowned when an iceberg capsized close to their umiak.[13]

The last opportunity for travel off the island of Umanak came at the end of October when Dr. Christensen, accompanied by her kivfak, took Isobel by motorboat to visit some of the outstations, clustered hamlets of earth houses scattered about the fjords and islands to the north and west of the colony. They visited Igdlorssuit on Unknown Island, where they stayed in the Greenlandic manager's house and Isobel had her first meal of raw *matak,* the skin of the white whale. She had previously eaten the milky-looking skin, with its layer of jellied fat, fried in bread crumbs or with curry sauce, but on this occasion she had to face it raw and found it tasted like india rubber.

From Igdlorssuit they crossed the strait to Upernivik Island and called at the tiny hamlet of Upernivik Naes, consisting of two or three sod huts. They were out in the strait again on their way home with the wind rising when the motor stopped and the boatman could not restart it. Fortunately the boat had sails, and with the help of a wheezing motor they were able to beat their way back to the coast they had just left. The Greenlandic cateket gave the party shelter in his igdlo. They entered the hut by an L-shaped earthen passage with a low door opening into a room with walls of unvarnished pine, unfurnished except for a tub and a pile of raw whale meat. A door led through to the sleeping room, containing two chairs and a stove in addition to the *brik,* or sleeping platform. Unexpected visitors in a community were the excuse for a party. Everyone crowded into the igdlo to share the doctor's food and listen to the cateket play jigs and reels on his concertina. Isobel contributed to the festivity with her "sole" accomplishment, dancing the sword dance. After grumbling about Isobel's display of Scottish patriotism, the doctor, not wishing to be outdone, danced the Charleston with her kivfak. At night all the women slept on the brik while the men occupied the outer room of the turf and stone hut.

The text of Isobel Hutchison's travel books, for the most part, follows closely the events recorded in her diaries and expanded on in her journals. In the Greenland journal she vented her exasperation with the doctor, the culmination of an already prickly relationship that is barely hinted at in her book. The problems began almost immediately on her arrival in Umanak. When Isobel was allotted a house of her own, it was agreed that she would continue what had been arranged and take lunch and dinner at the doctor's house, where the nurse lived as well. Within the small Danish community, the three unmarried women were known as the Three Muses. Accustomed to being the leading woman in the community, the doctor may have felt her position threatened by this exotic new arrival. Gregarious by nature, she also expected more of Isobel's time than Isobel wished to give. On the motorboat trip the doctor proved to be "a most disagreeable fellow traveller" who complained loudly about every inconvenience. Isobel, having found that her own kivfak could produce meals to her taste, wished to withdraw from the contract to take her meals with the doctor. She accomplished this gradually, buying her share of the doctor's stores. It is a tribute to Isobel's tact and diplomacy that she was able to maintain good relations while extracting herself from the doctor's sphere of influence. They continued to meet, attend parties in each other's homes, and share social occasions with the handful of Danish adults in the village. Open disagreement would have been obvious to everyone in the tightly knit community and would have created many awkward situations.[14]

The first snow fell on Umanak on October 19 as a fine dry powder, lightly covering the ground. It had the advantage of covering "the poisons of the undrained village" but effectively put an end to botanizing. Although there were few flowers to be seen at that time of year, Isobel—trained to identify the plants by their leaves and form—filled over three hundred envelopes with seeds from more than fifty varieties of plants during her first three weeks.

In her luggage Isobel brought not only the cine camera, projector, and screen but also several reels of movies. These included a series on the life of the Alaskan Eskimo, some films of Arab life, a staged version of the loss of an English submarine at naval maneuvers, and a comic American movie. At the pastor's request she put on a film show for the village at the schoolhouse, using sound provided by Scottish records on the gramophone, and to finish the evening Isobel performed her sword dance. The gramophone would have been a wind-up instrument, and it is possible that the handcranked projector was lit by a candle or lamp, since there was no electrical power. The evening was a huge success and was a clever way for the pastor to introduce Isobel to the Greenlanders of the community.

On the last day of October a surprising final delivery of mail arrived from the outside, coming north by motorboat from Disko Island. Even though the ship bringing the mail had left Copenhagen only eight days after her own ship, Isobel was grateful to receive a dozen letters. It would be March 15 before letters could go out from the colony, and then only if conditions were favorable. There were two radio receivers in Umanak, and each evening, if the reception was clear, the operator at Godhavn, with a more powerful receiver, sent out a bulletin of world news. The manager of the colony typed a copy of what he heard, and children carried the news sheet from house to house.

"Very wonderful is the coming and going of the sun in his lingering farewell to this land of the north." In that mountainous northern region the local people knew the dates in November when the last gleams of sunshine fell on each peak. By November 21 the direct beam of the sun no longer reached even the top of Umanak Mountain. Instead it was "bathed at dawn and sunset . . . in a theatrical Alpenglow like rouge upon a dead face."[15] From her windowed vantage point, Isobel was a captivated spectator to the drama of the skies, not only watching the fading light but also having a clear view of the moon and stars. In preparation for winter the village smith, a Danish-speaking Greenlander, spent a day putting up the storm windows, which somewhat obstructed the view. Coming from Scotland where the few hours of midwinter daylight are often gray and gloomy, Isobel was philosophical about the loss of the sun and considered it a novelty. But she was more concerned about losing even part of her dramatic view.

The sun would return on February 5. In the three months of darkness the social life of the colony left little time for reading. Almost any occasion was the excuse for a gathering, and birthdays were celebrated with much coffee drinking. Isobel had brought a stock of books, ranging from Hooker's *British Flora* and Chambers dictionary to the Bible, Shakespeare, Scott, Dickens, Robert Louis Stevenson, W. H. Hudson, and a Scottish history, which she readily lent to her Danish friends. A small lake nestled in the rocky hills above the colony provided a natural skating rink where a few Greenlanders shared wooden skates tied to their kamiker. Isobel and Niels Kristensen, the Danish carpenter, were the only two who had real skates.

The exercise provided Isobel with outdoor activity away from the stuffy stoves, but more pointedly, away from "the narrower life of the colony." That single comment is the only hint at the stifling, inbred nature of the tiny community. Although there were nearly three hundred Greenlanders in Umanak, there were no more than a dozen persons who fitted within the closely defined Danish-speaking circle. In the endless round of coffee parties, with no stimulus from outside, topics of conversation were continually recycled. Regular social gatherings with the attendant small talk, or *pularpok,* to use the Danish expression, were never a feature of Isobel's life in Scotland, and she adapted with a struggle.

Christmas and New Year produced a flurry of activity. Preparations began with Advent, four weeks before Christmas, with carol singing in the church and sewing and baking at fever pitch in the homes. Christmas trees were improvised from sticks covered with green ling, a heath plant that lay fresh beneath the snow. With the shop selling only provisions, Isobel delved into her trunks to find suitable Christmas gifts for everybody. Creativity and ingenuity were in the air.

Dorthe hired an assistant to help clean the house from top to bottom. Dorthe's own igdlo was scrubbed and polished before she entertained Isobel to coffee, a special occasion. In the traditional sod house, the guests sat on the wooden sleeping platform, the bedding neatly stacked in the corner and covered by a white sheet. Isobel enjoyed the warm and friendly atmosphere with members of Dorthe's extended family, who had been invited, and the pastor and his wife, who arrived by chance.

In the Christmas season, Isobel began to receive invitations to the homes of Greenlanders. Arriving early for coffee at the igdlo of the smith, she found him still in shirt and braces and his wife combing her hair. With only one room, "they received me just as graciously as a duke and duchess in 'full fig' at the top of their stairs could have done. . . . The Greenlanders have the pleasant custom of calling each other always by their Christian names . . . masters and servants . . . children and parents."[16] Coming from a society where class distinctions were the norm, Isobel found this equality refreshing.

She gave two parties. First, all the Danish people were invited on Boxing Day for haggis and flaming plum pudding, followed by music and games. Dorthe served from the kitchen and then joined the guests for the games. In Dorthe's estimation a party of Greenlanders would have gone more heartily. In return for all the hospitality she had received, Isobel gave an afternoon *kaffeemik* (coffee party) for Greenlanders, with Dorthe in charge of the invitation list. The party was judged a great success: all the food and cigarettes were much enjoyed, and photographs and games of checkers took the place of conversation in a difficult language.

Isobel's party for the Greenlanders was in striking contrast to the holiday parties given farther north in Greenland in 1892 by Robert Peary, the polar explorer, and his wife Josephine. After serving a traditional meal to the men of his expedition, the Pearys gave a separate Christmas dinner for the Greenlanders. "It was amusing to see the queer-looking creatures, dressed entirely in the skins of animals, seated at the table trying to act like civilized people." On New Year's Eve Mrs. Peary, dressed in a black silk tea gown with a canary silk front trimmed with black lace, gave an "at home" for the men of their party, serving chocolate ice cream, cake, and crullers she had made herself.[17]

The circumstances experienced by Isobel Hutchison and by Mrs. Peary were markedly different. The Pearys were living beyond the boundary of the Danish colony, among Greenlanders who had rarely if ever seen a white person. The population of Umanak had been under Danish administration for about two hundred years and had absorbed some Danish culture. Whereas Isobel was clearly intent on adapting to local customs and entertaining the native people as her friends, Mrs. Peary, in the North as wife of the explorer, was attempting to recreate southern conditions to please her husband. Conscious of his status as leader of the expedition, Peary believed he was bringing the benefits of his culture to a primitive people.

Throughout Greenland there are many people of mixed blood, as a result of the long history of contact beginning with the whalers and the explorers. There is unconscious racism in the patronizing tone of the official description of the population:

> During the early part of colonization the persons appointed at the various trading posts were almost exclusively Europeans. The majority of these employees remained in Greenland and married native women. The resulting mixture of native and European blood has proved physically and intellectually superior to the pure Eskimo and more susceptible to development. From the middle of the nineteenth century the Danish administration increasingly began to hire natives and at present all the *subordinate* [my italics] officials as well as the *lower* clergy and catechists [teachers] are Greenlanders.[18]

In the twentieth century, as transportation became easier, the young administrators often returned to Denmark to marry and brought their wives to the colony, sometimes choosing to ignore their previous Greenland alliances and their mixed children. Danes who married native women might be ostracized from Danish society in the colony.[19] There was little or no social intercourse between the Danish ruling class and the native people. People of mixed blood were on the fringe of Danish society, depending on their adaptation to the culture and their facility with the language. At Umanak, throughout the Christmas season the church was filled to capacity with Greenlanders, but only at the special Christmas afternoon carol service in both languages did the ten or twelve Danes of the colony attend. Pastor Rosing and his Greenlandic wife, Sara, were accepted because of his position in the community, but when the Rosings entertained they did not mix the two groups, although they usually included Isobel in their parties for Greenlanders.

In the diary of a young Danish governess to the children of the manager in Umanak in 1899, the segregation of the races is evident by what is not recorded. Lisbeth—surname unknown—described the social life of the colony very much as Isobel saw it thirty years later, with one significant difference. She made not a single reference to a native Greenlander. The presence of native servants in the house was taken for granted, but they were never mentioned. The fact is that no Danish household could have existed without them. Lisbeth seemed to be living on a beautiful exotic island peopled entirely by Danes, enjoying Danish society and Danish customs.[20]

Whether she liked the racial situation or not, Isobel could do nothing to change an attitude that was so ingrained. As a Scot she was fortunate to be accepted by both groups. She appreciated the kindness of the Danes, but her real interest was in the Greenlanders, and she persisted with their difficult language in order to learn more about their culture. As with her reaction to children, she did not instantly profess to like all native people, but her affection for individuals grew as she became better acquainted with each of them.

Isobel regularly attended services in the Greenlandic language. "When I go to church with Dorthe, carrying the hymn-book which Fru Rosing has kindly lent to me, I feel like some dignified Scottish gentlewoman of last century accompanied by her body-servant. But when we climb again to the Baadsmand's house and sit together on the kitchen dresser drinking cocoa, I lose this feeling."[21] It was not part of the agreement for Isobel to provide Dorthe with meals, but it soon became the regular arrangement. Dorthe was more than a kivfak; she was becoming a friend and confidante. The feeling was shared. In Dorthe's words, "Tuluk—never angry!" Tuluk, the word for biscuit, a favorite commodity since the time of the Scottish whalers, was the name Isobel was known by in the colony.

Although she adapted cheerfully to the Greenlandic diet of seal, whale, or whatever was available, Isobel sometimes hankered for familiar Scottish fare. On those occasions she invaded the kitchen and tried to construct dishes from imagination and memory, usually with less than successful results. Without Dorthe to cook and look after the house, her life would have been nearly impossible. In fact, one day when Dorthe was ill and suggested she might not be able to work the next day, Isobel found the idea too terrible to contemplate. Dorthe turned up even with a fever.

Typhoid was one of the constant fears in the colony because of the lack of sewerage. Isobel did not mention the sanitary arrangements, but her statement regarding dogs speaks volumes: "In this typhoid-ridden colony, where the dogs are the scavengers of the houses and devour literally everything that can be devoured, they must be handled with caution."[22] That remark, in addition to the fact that the village was undrained, creates an unpleasant picture. Consumption was another scourge, and the carpenter was building a new sanatorium separate from the hospital.

For the local people, Isobel was a new source of income. Prices of commodities such as freshly caught fish or seal were extremely low by British standards, and she was pleased to buy whatever came to the door. As the winter progressed and natural shortages occurred, Greenlanders sometimes came to "borrow" money. She devised ways of putting money into the economy by paying people to pose for her to draw or paint or by buying their drawings or handmade artifacts. Such transactions brought a growing stream of visitors to the kitchen door with less and less desirable items to sell. To counteract this, she arranged to take only a certain amount of money out of the Danish manager's bank each month so she could point to her empty purse when it was gone. If Dorthe was present when people came importuning her, the great protector had no hesitation in sending them away.

Coming from Britain, where pets are as privileged as children, Isobel found the treatment of dogs in Greenland hard to bear. At her home in Carlowrie there is a plot of ground beyond the kitchen garden where pets were laid to rest, their names engraved on a headstone set into the wall. At Umanak the carpenter had six fine dogs that he planned to use for sledging when the time was right. Joko, who looked like a black-and-white Scottish collie, was the favorite and became Isobel's pet, often accompanying her skating and always turning up at the door for dog biscuits. By contrast, most of the native dogs were miserable and neglected, scavenging for food. Greenlanders, working desperately to feed themselves and their families, could not afford to feed dogs unless they were being used for sledging. When it was necessary to kill dogs they were hanged, ammunition being too precious to expend on them. The women used dog skins to make soles and linings for winter kamiker.

Although in 1928 snow fell heavily between Christmas and New Year, ice did not form in the bay—a misfortune for the colony, which depended on fishing for halibut and catching seals on the sea ice. In recent years it had become the exception rather than the rule for Umanak fjord to freeze. Many dogs were hanged for food, and their agonized howls rent the air. Even Rasmussen and Mikkelsen, whose expeditions depended on their faithful dogs, had to sacrifice them one by one in dire circumstances to feed either the remaining dogs or themselves.[23]

By mid-February the thermometer slipped to 10°F and held long enough for the ice to set solidly in the bay. The people, who had been held prisoner on the tiny island since October, began to drive dog teams cautiously over the ice to fish and hunt. Isobel joined two fishermen going far out on the ice floes, and looking down through a fishing hole into the clear water, she was inspired to poetry:

Watching the Sea-Floor

To-day I saw the strange seaflowers
That bloom on ocean's floor,
Stretched on the ice I lay and looked
Down through the fisher's door.
I saw the kanajoq spread his wing—
Steered by his spotty tail—
And once I saw a cod slip past
Like a nun in dim grey veil.

I saw the slippery weeds that blow
Deep in that garden-place
As they waft their fingers to and fro
In the currents' noiseless race;
The water was bright with the crystal light
Of the covering ice and snow,
And I thought of the wonderful Heart of God
As I gazed on that life below.

I thought of the glorious Hand that wove,
The glorious Love that planned,
And I wondered why God set this garden-grove
So deep out of sight of the land.
Was it but that the fish with his round bright eye
Should joy in its forests rare,
And pass in the silvery brightness by
Of His love still unaware?

I thought as I lay on the frozen sea
And peered on that stranger shore,
How maybe some other looked down on me
Through a rift in God's crystal floor,

And wondered and pitied to think that I
(A-swim in earth's sleepy brine)
Was as much unaware of His golden eye
As the fish unaware of mine.

As daylight returned the sun strengthened quickly, and by April that fixture of the winter, the train-oil lamp, was no longer needed. This was the signal for Dorthe to hire her friend to help wash the parlor literally from floor to ceiling, the wood having been thoroughly blackened by smoke over the winter. Isobel donned an overall and joined the work party. Can this be the same Isobel who, on her arrival in Umanak, described Dorthe as the most "ornamental parlour-maid" she would ever "possess"? It is an indication of her evolution as a fully realized person.

The mild winter meant the early breakup of the ice, and ships were already sailing to Godthaab from Denmark. The baadsmand was on his way back to the colony with his new bride, and Isobel would have to vacate his house. The question of where to live was quickly settled by the pastor, who rented her his studio in the loft of the *praestebolig,* and Dorthe could continue there as her kivfak.

As winter ebbed and flowed, cold weather returned in mid-April, producing sea ice strong enough for travel. Isobel, accompanied by a Greenlander, borrowed the carpenter's dog team to cross the fjord to Nugssuaq peninsula several times for picnics, sketching, and seed collecting. But as quickly as the ice had come it was gone, and once more they were prisoners on the island rock.

In seven months Isobel became intimately familiar with the life of a tight little northern community. Inevitably, underlying tensions existed, but they were camouflaged by careful etiquette. Greenlanders and Danes alike coped with the months of darkness by celebrating every possible birthday with a kaffeemik or, in the case of someone important like the Danish manager, a dinner party. The endless round of meaningless social gatherings emphasized the narrow life of the colony to Isobel, who would have preferred to spend more time writing, painting, studying, and in outdoor exercise. Her interactions with the native Greenlanders were paramount to her, and she attended their important occasions—christenings, weddings, and funerals in the church. While she frequently mentioned activities with the doctor, the nurse, and the carpenter, and occasionally with the family of the manager, her friendship with Otto Rosing and his wife continuously provided a link to the native population and a supportive home relationship.

In describing the domestic arrangements, both her own and those of some of the Greenlanders, Isobel was following the style of other women travel writers. The painted wooden houses of the Danes, with their pointed rooflines and casement windows, contrasted with the flat-topped earthen igdlos, built of clods of turf set with large stones. Some of the igdlos were lined with unvarnished pine supplied from Denmark, some with Danish newspapers or with pages of the *Illustrated London News*. Food has an important place in her narrative, both the native items to which Isobel readily adapted, such as matak, seal-liver pâté, seagulls' eggs, and blood pudding made from the blood of the white whale and Dorthe's attempts at familiar Scottish fare. Even the meager monthly grocery list is of interest, containing such items as *kaffina,* used to stretch the regular coffee, dried potatoes, coal, and Christmas candles. Although she only hinted at the unpleasant lack of any form of sewerage, it can be deduced that everything, from the dishwater to the contents of the chamber pots, was thrown out the back door onto the rocks for the scavenging dogs. Such details are not found in the heroic narratives of male Greenland explorers, and they add a valuable element of social discourse to her story.

The time of release from the island prison was fast approaching. From the late October motorboat trip with the doctor to the outlying communities until the brief time in late winter when the sea ice permitted travel by dog team, everyone had been confined to the island. Despite the magnificence of the surrounding scenery, the village, huddled on the only relatively flat portion of an immense rock, had little scope for the long solitary walks Isobel loved. Having spent the months in a small village where one's every move was noted, she was ready to expand her horizons. With the navigable open water of spring, the opportunity was now at hand.

6

Unknown Island

Before moving to the studio loft of the Rosing house, Isobel—with Dorthe—gave a party for forty Greenland friends, with coffee and dancing, as a farewell to "Scotland Yard," her home of the past seven months. With advancing spring and long hours of daylight, she spent more and more time out of doors, sketching, painting, and watching for the first spring flowers.

There was great excitement in the pastor's house one morning. Otto Rosing was out in his kayak with three other kayak men when white whales were sighted and one was killed. This brought the whole village down to the beach to share in the matak, cut and eaten on the spot. Isobel filmed the scene with her cine camera. Such bounty was a boon to the whole village, since the whale contained a perfectly formed baby whale, convertible into cash.

With the arrival of open water, the time had come for the pastor's annual visitation to the little communities around Umanak fjord, and Isobel was invited to go along. She would be able to spend time botanizing while the pastor collected young confirmands and took them back to Umanak for the special service at the end of June. The strong wooden sailing boat with eight rowers, accompanied as always by a kayak for safety, was dwarfed by the wide waters of Umanak fjord, overhung by soaring glaciers.

Thus began a phase of Isobel's year in Greenland that suited her better than the sociable confines of the village. She would be traveling with men employed on a serious task; meals and accommodations would be improvised as needed. Her command of Greenlandic was now sufficient that she could leave the group and stay alone in isolated communities to search for plants or sketch. At last her independent nature could be given full rein. Her simple words "We are out and away at last!" suggest her anticipation and excitement on being released from the "island prison."

The men in the crew were some of those Isobel had come to know and admire during the past six months. Andreas, the kayak man, was "a giant of a man, big in all ways, the best husband as well as the best hunter in the colony." Lars, the steersman, was "a great, kindly, jolly fellow . . . one of the best step-dancers . . . whose bright brown eyes, sharp as a bird's, search the horizon for ice." Her friendship with Ole, the bell ringer, began on the initial October picnic. Tittus, the artist, was a simple man, treated like a child by the other men although he was the oldest. Another was Aboru, "whose round flabby face is the 'home of the happy smile.' "

After crossing the fjord to the mainland, Isobel stayed for a week at the home of the Danish manager at Qaersut on the north side of Nugssuaq while the pastor continued around the tip of the peninsula. Although it was still bitterly cold in the last week of May, on that side of the fjord the flowers were beginning to sprout and the leaves of the ground willow were turning green under the snow.

When Isobel rejoined the boat the tour continued to the outstations of the inner fjords. Sometimes they landed in the dead of night, still bright under the midnight sun. Their arrival was the signal for an early morning communion service in the cateket's house if there was no church, or in the pastoral tent spread out on the shore. Rosing performed marriages, baptisms, and school inspections, since all Greenland schools were under the supervision of the church and the catekets were the teachers. Because the rowers were each paid two shillings a day, the tour often continued without a night's pause to take full advantage of favorable winds. They often traveled as much as forty miles a day.

At Ikerasak Island, tents were set up in a meadow where the vegetation was further along than at Umanak, spring being more advanced as they went deeper into the fjord. Isobel was summoned in the evening to drink coffee in one of the village homes.

> As I sat, more and more Greenlanders poured into the little room, till I was surrounded by a gazing crowd who watched my every motion. It was the first time they had ever seen a "Tuluk," and I should have been glad to know what impression the strange sight created, for a strange sight I must have

been! The weather was cold, and over my knickerbocker suit and sealskin kamiker I wore a kilt of the Macleod tartan, whose thick pleats I found admirable for keeping in the warmth. Above it I wore the tweed jacket of my suit, and on my head was squashed a grey felt hat with a wide brim to shield my eyes from the glitter of sun on ice. As my conversation was strictly limited to such matters as asking the names of the infants and inquiring about the weather, I am still in ignorance of what the ladies of Greenland thought of the kilt![1]

On this trip she learned to eat whenever and whatever the occasion offered. She went off to sleep that night in her tent to the sound of drums and slept soundly despite the bitter cold and light snow drifting down.

In the passage quoted above, Isobel, in poking fun at herself in hybrid costume of breeches, kamiker, and kilt, employs a device frequently used by writers of the period to signal the novelty of their travels. In a humorous way she is pointing out that she is participating in the unique experience of being observed by people who have never seen a Scotswoman before. She is simultaneously the observer and the person being examined in detail.[2]

As they rowed on the next day, the whole crew, including the pastor, was distracted by the sighting of whales. The kayak man and Rosing, who had also brought along his kayak, both gave chase. The large boat put ashore on a rocky headland, where Isobel and the crew brewed coffee and waited three or four hours for the hunters to return with their catch, a small white whale. The tent was set up for shelter while some of the meat was cooked. Rosing and the crew eagerly devoured the gray matak raw, but Isobel cooked hers in butter. It was by then six in the evening, and they were only an hour from their starting point with a six-hour journey ahead before they reached the large community of Satut.

They had crossed the strait and passed under the foot of gigantic precipices three or four thousand feet high, the home of colonies of seabirds. Every year Greenlanders risked their lives collecting birds' eggs—adding welcome variety to a diet primarily based on seal, whale, and fish. Isobel was as impressed by the dangers Greenlanders would accept in gathering food as she was awed by the grandeur of the scenery. She fully understood the risks regularly assumed in the hunt, whether harpooning a whale from a delicately balanced kayak or going out onto the treacherous sea ice to fish for halibut.

But while Isobel accepted the hunters' instinct to kill wildlife for food, she was distressed by their disregard for the victims, especially when wounded birds were left to die in the bottom of the boat or swung flapping from a broken wing to decoy other birds closer. When a fusillade of guns broke the peace on a perfect Sunday morning, it was more than she could bear. She exploded in anger, denouncing even the pastor for the cruelty she had witnessed.

That same evening Otto Rosing, the hunter transformed, wearing black gown and white ruff, conducted the communion service in the pastoral tent using packing cases as an impromptu altar for the handful of natives living in the remote community. It was the last communion service of the tour and at the same time the simplest and grandest. Such contrasts were the essence of Greenland life.

At the end of the tour Isobel was landed on Ubekendt Ejland. It had been arranged that she could remain at Igdlorssuit on this "Unknown Island" for a month, occupying a little Danish wooden house that was not being used by the Greenlandic manager for whom it was intended. Margrete, sister of the manager, Johan Lange, acted as her new kivfak. "I am very happy in my new home. The truth is, I am glad to be alone for a little; glad to be free of the hospitable round of coffee parties and 'pularpoking' which take up so much time in our island capital, though I do not lack for visitors in my new retreat."[3] Her command of Greenlandic was now sufficient for her to manage without Danish interpreters. For the rest of her stay in Greenland she would be using mainly that language.

The scene from the window of her house, out across the wide fjord dotted with icebergs to the snowcapped mountains of the distant peninsula, was a constant joy to Isobel. Similarly, the beauty of Greenland captivated the American artist and writer Rockwell Kent. Kent, shipwrecked in southern Greenland in 1930 after sailing in a small boat from New England with two young men, was granted permission to stay on and paint until the last boat of the season returned him to Denmark. He returned the following year with the material to build a house at Igdlorssuit, three years after Isobel spent her month there. His colorful chronicle of the year on Unknown Island, *Salamina,* is named for the kivfak who ruled his house.

In the manager's motorboat Isobel went on botanical excursions to the neighboring islands of Karrat and Qeqertarssuaq. She hoped to reach the peninsula of Svartenhuk, where there was rumored to be a mysterious human fossil, seen some years before on the bank of one of the rivers. The weather suddenly turned cold, and she and the boatman camped on a small island, sleeping under a blanket of snow. On reaching Qeqertarssuaq they were told they would need a dogsled to reach their destination on the peninsula. German mineralogists to whom she later gave the information were unsuccessful in their hunt for the fossil.

On Ubekendt Ejland, formed from basaltic rock, the flora varied slightly from that on Umanak and the interior islands of the fjord. She could find no record of the island's being explored botanically, and in the month she was there she collected and mounted many specimens, particularly of saxifrage, which grew in great variety.

Isobel hired the manager's motorboat and made a circular tour of the island, which has no roads, even swimming from a sandy beach beside the icebergs. By the time she was ready to leave Igdlorssuit she had made many friends, including Samuel Moller, the cateket, and his wife, who were kindly and "disinterested"—a trait appreciated by Isobel, who cherished privacy. Two others were Scarita and her effeminate son Paul, an anomaly among Greenland men, who did his own sewing, knitting, and crocheting. Although she wrote less about it than she wrote about Umanak, the month on Ubekendt Ejland was possibly the happiest part of Isobel's Greenland year. She was content to use the time writing, collecting, exploring, and taking the solitary walks not possible on Umanak.

From Ubekendt Ejland, Isobel watched the snows melting from the mountain of Qilertinguit, on the Nugssuaq peninsula, at 6,250 feet one of the highest peaks in northwestern Greenland. During the winter on Umanak the mountain had stood like a beacon. It had last been scaled in 1879 by the Danish explorer R. J. V. Steenstrup with three Greenlanders, and before that by the English artist Edward Whymper in 1872. Returning to Umanak in the motorboat, Isobel immediately embarked on her goal of climbing the mountain.

On her way to the mountain, Isobel picked up Kruse, the pastor's assistant at Umanak, and they landed at Qaersut, an outstation on the Nugssuaq peninsula. She had arranged for Thomas Nielsen, who hunted on the slopes of Qilertinguit, to accompany them on the climb. Thomas was reputed to be the most reliable man in Qaersut, the foremost hunter, and in the eyes of the frugal Greenlanders, well-to-do. The starting point was to be the beach at Sarfarfak, where Thomas had an igdlo.

The mountain was veiled in mist during the three days they waited in bitter winds for the weather to clear. Although it appeared no better to Isobel, the men recognized signs of improvement for the following day, and they set out for Sarfarfak. When they arrived Isobel, still despondent about the weather, set up her tent, camp bed, and sleeping bag beside the igdlo where the men slept.

The next morning, although the mist hung as low as ever, Thomas prophesied that it would clear and said they should set off before the day became too hot. At 10:40 A.M., carrying a kettle and some food, the three climbers started up the little river valley. As they went round the shoulder of the mountain to the southern face, gradually the mists began to clear. Isobel was fortunate that her two guides were nearly as keen to reach the summit as she was. Her experience with Greenlanders was that they were easily discouraged and understandably cautious about wind and weather, unlike herself, who was rarely dissuaded from pursuing any goal.

They stopped at four in the afternoon to make tea, the last place they would find fuel. After a stiff climb up slopes of loose shale, sometimes covered with snow, they could see the peaks of Disko Island and Davis Strait far to the south. At 9:00 P.M. they could see the summit of the mountain not far above them. To reach it they needed to use the rope they had brought for the only rock climbing involved. The summit, gained at last, was a long, flat ridge ending abruptly in a sheer inward-sloping precipice leading straight down to Qaersut. At the summit was a cairn containing a schnapps bottle holding a paper with Steenstrup's signature along with the names of his companions. They added their own names to the bottle along with a small Union Jack that Isobel had brought.

In the bitter cold at the summit they stayed only forty-five minutes, starting down at 10:15 on "a most lovely evening, the mists entirely gone, the stillness of the approaching Sunday dawn already breath[ing] upon the snows." After a picnic tea on the slopes at 3:00 A.M., they reached the hut at Sarfarfak feeling fresher than when they had started. It was a radiant morning, and they were luckier than they knew—those two days in mid-July were the only clear ones between weeks of rain and mist. Isobel had been "astonished and delighted" as they neared the summit and was altogether exhilarated by the success of the climb. She listened to the excited account her Greenlandic companions gave to the boatmen waiting at the bottom before tumbling happily into the camp bed in her tent.

By arrangement Kruse returned to Umanak and Isobel stayed on with Thomas and his family to attempt a trip to the end of the Nugssuaq peninsula. Using a rowboat this trip should have been a matter of three or four days, but the weather was against them from the start. Both Isobel and Thomas were suffering from severe colds, the aftermath of their climb. After waiting all day for the weather to improve, Isobel, Thomas, three young men, and two women set off but were caught in a downpour and had to put in at Ikorfat, where Thomas had another sod house. After supper there, Isobel retired to her tent on the shingle above the beach, but she awoke about 4:30 A.M. in a howling gale that whipped the tent pegs loose as if they were matchsticks. She was obliged to go into the igdlo to wait out the storm. High seas held them at Ikorfat for a second night, and Isobel moved into an abandoned sod hut, even dirtier than the one occupied by Thomas and the crew.

Scarcity of provisions motivated the party to launch the boat on the now calmer seas, and they reached Nioqornat, twelve miles along the peninsula, where there was an outpost. Isobel hoped to continue around the headland to visit the grave of a Scottish whaler buried there in 1825. She was always conscious of her Scottish roots and used any occasion to remind her friends of them. Possibly she was trying to avoid being totally submerged by a foreign culture, or perhaps she used this device to hide traces of homesickness.

Torrents of rain descended over Nioqornat for three days. As always, the arrival of visitors at an outpost was the signal for entertainment. Word of Isobel's sword dance had preceded her, and she was persuaded to perform while Hans Petersen, manager of the outpost, a southern Greenlander, accompanied her on the concertina.

> Next night, I was summoned to the kitchen to see a delightful old woman, Anna (aged sixty-three), who, I was told, could give an excellent performance of the Sword Dance just as she had seen me do it! And so she did, with her arms in the air and her sallow little face all wrinkled into smiles and her toes nearly coming out of her tattered kamiker. It was the most comical

performance! Though the steps were wrong, no matter! Anna got round in fine style, and was such a gay, cheery little soul withal that it did one good to look at her.[4]

Isobel's vivid description of Anna is both admiring and mildly condescending. Although the well-bred Edinburgh lady would shrink from giving offense in her writing, she saw her carefully learned performance unconsciously parodied by a ragged old lady of spirit and intelligence. The moment encapsulates many of Isobel's contradictory reactions to the Greenlandic people. She admired their cheerful acceptance of the harsh conditions of their lives, yet her Scottish pragmatism was sometimes impatient with their seemingly lackadaisical attitudes. In a land where time meant little she could be irritated when they refused to be hurried, yet underneath she understood that they were molded by their environment. Coming from a home where she lacked for nothing, she admired people who were happy with so few material comforts.

Isobel and Hans Petersen sketched Anna the following day and listened to her memories of contacts with the Scottish whalers earlier in the century. The whalers, from the large British fleets, hunted in Davis Strait off the west coast of Greenland from the mid-eighteenth century until the market for whalebone and whale oil collapsed at the beginning of the twentieth century. Typically the ships left Britain in late February and were gone no longer than five months. Each ship was manned by sailors from its home port, and they were a closely knit group of men, handpicked by the skipper. Contact with Greenlanders was limited, since the Danes permitted access to only three harbors for filling water barrels.[5] The legacy of contact with the whalers in Greenland was far less damaging to the native people than what Isobel was to see on her future travels.

With the ship to Denmark expected sometime in August 1929, Isobel had to abandon the attempt to reach the end of the peninsula. In stormy seas she, Thomas, and their crew turned back and were glad to reach the sheltering turf huts at Ikorfat once more. This time Isobel pitched her tent inside the filthy abandoned hut to keep off some of the rain leaking through the turf roof. As the bad weather continued, the group hunkered down and amused themselves with checkers and acrobatics. Isobel paid each of them to act as a model for her sketches.

With the sudden appearance of her friends Otto Rosing, Kruse, and Samuel Moller from Ubekendt Ejland in the doctor's motorboat, Isobel snatched up her camp bed and joined them. They were going around the peninsula to a meeting, and they could drop her off on the headland at Nugssuaq to be picked up on their return. At Nugssuaq Isobel searched for plants and then set up her camp bed in the shelter and privacy of the church to recover from her exertions. It had been nearly three weeks since she left Umanak to climb Qilertinguit. In the company of a variety of Greenlanders, she had battled impossible weather and now finally reached her goal, the grave of the Scottish whaler at Nugssuaq.

What are we to understand of Isobel's tenacious obsession with the grave of the Scottish whaler? Not long after her arrival in Umanak on her first visit to the Rosings, she had seen a sketch Otto Rosing had made of the grave marker and inscription. The death in 1825 of a Scotsman of about her own age on that lonely shore far from home had so captured her imagination that she pictured the scene of the burial, the service read by the ship's master, and the marker carefully incised by the dead man's friend. Thoughts about death, harking back to the deaths of her brothers, still lingered in the corners of her mind. Knowing that the grave was rarely visited, she felt a duty to pay her respects to her fellow countryman. When Isobel decided to do anything, it took more than bad weather to stop her. On a misty evening the boatmen put her ashore on the wild headland, and one of them climbed up to the grave with her. She read the inscription, picked a willow twig to root at home in Scotland in memory of the whaler, then left him to his rest.

Only a few days after Isobel returned to Umanak from her latest adventures, the *Hans Egede* sailed quietly into the harbor. On a cold day late in August 1929, the ship was ready to steam out with Isobel and the manager's family among the thirteen passengers. There were final cups of coffee and affectionate farewells, especially for Dorthe, who had taken such good care of her. *Inuvdluarna!* Farewell—shall we ever meet again? Isobel's long Greenland adventure was over.

It was two years since Isobel Hutchison had first set foot in Greenland, and she had been greatly changed by all she had experienced. On her first camping trip to the birch grove she referred to one of the women as her "lady's maid," and on arriving in Umanak she had described Dorthe as "an ornamental . . . parlour-maid." These phrases came naturally to one who was raised with servants to cater to every need. Isobel soon learned the democratic nature of Greenland society, where the kivfak rules the household. Rather than being served by Dorthe in solitary state, she preferred that they share their meals. Toward the end of her stay she even helped clean the house, working along with Dorthe and a friend hired to do the job. Their relationship became one of friends rather than mistress and servant.

The fastidious person who in the beginning needed a room to herself and required a hot bath each morning became the accepting traveler who could bed down in tent or hovel and sleep soundly, whether she was alone or in a room full of people. Whenever possible, however, she would choose privacy. Although she wrote very little about it, the month spent alone on Ubekendt Ejland was perhaps the happiest part of her whole Greenland adventure. As with accommodations, she accepted the unfamiliar diet of seal, whale, and seabirds without complaint. Although she had a close affinity with animals and birds, she never questioned the hunters' need to kill for food, only protesting unnecessary cruelty.

The two visits to Greenland were adventures comparable to those of two famous Victorians: Mary Kingsley's risky travels with tribesmen in West Africa and Isabella Bird's daring ascent on horseback of a Hawaiian volcano.[6] Greenlanders, the most hospitable of peoples, presented no threat, but Isobel faced physical danger whenever she ventured off the island of Umanak in an umiak or a rowboat. Greenlanders, mainly men but occasionally women, regularly faced that same danger. In choosing to accept their ways of travel Isobel distanced herself from the Danish women in Greenland. The native people, accustomed to white women's being part of the ruling establishment, gradually accepted her as a friend as she showed her willingness to participate in their life and struggled to learn the rudiments of their language.

Isobel Hutchison had spent the year among total strangers on a tiny island within a closed society, speaking Danish and learning to get along in Greenlandic, which Fridtjof Nansen referred to as a "formidable" language. In Nansen's words, "One cannot help being comfortable in these people's society. Their innocent, careless ways, their humble contentment with life as it is, and their kindness are very catching."[7] Isobel received hospitality and kindness from Danes and Greenlanders alike, and she always tried to give back at least as much as she received. Her annual resolutions listed her determination to enjoy every experience, to be generous, and never to be afraid of anything or anyone. In Greenland she lived those resolves, and so opportunities were offered to her that might have been withheld had she been timorous, stingy, or lacking in imagination.

Greenland, closed to tourists, was visited by few persons whose experiences were comparable to hers. Rockwell Kent's year on Unknown Island, described in *Salamina,* comes closest, although as a man he was able to control events in a way not possible for a woman. For example, he went on difficult and dangerous winter journeys, carried on a feud with the Danish manager in the community, was boisterous and convivial, and for the last part of the sojourn was joined by his wife from the United States.

Twenty years after Isobel's stay, a Finnish woman, Inga Ehrstrom, spent a year on Umanak, which she described in *Doctor's Wife in Greenland* as "a primitive community to which tourists have no access." Ehrstrom and her husband, along with their two children, were there to study the psychological effects of the climate on the native people. Attractive and high-spirited, Ehrstrom had a tendency to step outside the rules of the community—something the plainer and more conservative Isobel would never do. Ehrstrom found the Greenlanders "the simple and natural answer to a puzzle. A creator's intention unspoilt; an instrument moved by the wind and by desire. They seemed in their naturalness in some way to be nearer the meaning of life than we."[8]

After her 1927 trip Isobel Hutchison had produced a manuscript based on her visit to the east and south coasts, but it was rejected by one publisher as "attractive but not financially viable."[9] With the additional year in northern Greenland, including a winter, the material had sufficient novelty as a travel book to be accepted for publication. *On Greenland's Closed Shore,* was issued a year after Isobel arrived home.

Her book is based so directly on her detailed journals that one can watch her progress in adapting to Greenland life. This is particularly true in the section dealing with the year in Umanak. As she becomes more fluent in the language, Greenlandic words are increasingly sprinkled through the text. The need to invoke Scottish references decreases as she becomes more integrated into the Greenland community, although it never disappears altogether.

The review in the *Times Literary Supplement,* November 13, 1930, noted the writer's enviable facility but suggested that had she had more difficulty in finding the words to express herself the book might have been improved! Word pictures and allusions, literary or otherwise, flow so freely from her pen that one must read between the lines to appreciate her adventurous nature. She never draws attention to difficulty or danger, reporting her travels as though they were everyday occurrences.

In the preface to *On Greenland's Closed Shore,* Knud Rasmussen praised her for her knowledge of Greenland's history, for her humility, for her deep interest in the people, and for telling the story of her experiences with humor, sympathy, and delicacy of feeling.

Had Isobel been content to spend her year within the village of Umanak, her experience would have been limited to Greenlanders with a veneer of Danish culture. By spending the final months in outlying parts of the fjord, she had met many Greenlanders on their own terms.

Greenland, closed to casual tourists and outside enterprise, was a country where "the man of means could starve . . . in the midst of plenty. He can't demand a thing, for Government is not of but for a people of whom the white-faced traveler is not a part."[10] In other words, having granted permission to enter the country, the government took no further responsibility for the visitor. It was a different situation in the north of Canada, as we will see in a later chapter.

7

Prelude to Adventure

Although Isobel had written much of the manuscript for *On Greenland's Closed Shore* before she arrived home in September 1929, it would be late the following year before the book was published. To demonstrate her seriousness as a plant collector, she included an appendix listing in proper botanical form the 196 plants she had found during the two visits to Greenland and giving added details on seeds that had since germinated in Britain. Isobel undertook the identification using a book on Greenland flora, but she also consulted Dr. Morten Porsild and his son, Erling, of the Arctic Research Station on Disko Island. The complexities of spelling Greenlandic words and place-names in the text gave the publisher problems. Isobel's attention to detail can be seen in her handwritten corrections of errors that had escaped the copyeditor, which appear in some copies of the book.

In addition to her book, Isobel produced articles on Greenland and Iceland, gave a talk for the BBC, and was entrusted by Knud Rasmussen with translating his book of Alaskan Eskimo tales from Danish into English. This was published as a deluxe illustrated edition in New York in 1932 under the title *The Eagle's Gift*.

Long before *On Greenland's Closed Shore* appeared, Rasmussen gave Isobel a glowing recommendation as a lecturer on Greenland:

> I have met Miss Hutchison in Greenland and there I have had the oppor-
> tunity of noticing the interest and profoundness with which she made
> herself familiar with the life and customs of the country. . . . Few foreign
> travellers have seen so much of Greenland as Miss Hutchison, and as she,
> into the bargain, is an excellent and keen observer she will certainly be a

greatly requested lecturer, especially in England where the knowledge of the conditions in Greenland is rather small.[1]

With her Greenlandic costume and wide selection of slides, cine film, and artifacts, Isobel was soon in demand as a speaker. On returning from a trip abroad, Isobel always showed her slides and gave a talk to the women's guild of her church in Kirkliston, but her fame now spread far beyond home territory. She lectured all over Scotland, to scientific groups, to the Royal Scottish Geographical Society, and to women's clubs. Sometimes she was paid an honorarium, sometimes only expenses, but many times she spoke gratuitously.[2]

After fulfilling all her obligations, Isobel accepted no further speaking engagements after the end of March 1931. Her mother was seriously ill. Isobel and Hilda hired a nurse and took turns sitting with the patient at night. Nita arrived for a visit from the south of England, where her husband had retired from the Royal Navy. A second nurse was employed, and for a few weeks Mrs. Hutchison seemed to rally. The doctor ordered champagne for the invalid, and the chauffeur took Isobel to Leith to get it from the family wholesale business. Mrs. Hutchison weakened steadily yet continued to live until the evening of June 15, 1931. She was buried three days later beside her husband and two sons, under the Celtic cross in the churchyard of the ancient kirk of Kirkliston.[3]

Death, the subject underlying many of Isobel's deeper thoughts, had come slowly and naturally to her mother, in contrast to the deaths of her father and brothers. Having been confined to home and sickroom for weeks on end, Isobel and Hilda left instructions for the house to be cleaned from top to bottom and went on holiday to Ullapool for two weeks of walking in the glorious hills of the west coast of Scotland.

Isobel was now head of the household at Carlowrie, and decisions about the future needed to be made. On their return, the estate was inspected by an appraiser with a view to putting it up for sale as recommended by the trust lawyer who managed the estate. It was decided to delay putting it on the market until after the general election late in October, and at the same time the two sisters considered the option of leasing the house.[4] Isobel had tasted freedom and had a vision of new horizons to explore, and neither she nor Hilda, a gifted musician with travels of her own to pursue, was sure she wanted to be tied to Carlowrie despite their deep attachment to the place where they were born. Perhaps because of the state of the market at the beginning of the Great Depression, no sale occurred, with unfortunate consequences much later. The house was leased for eighteen months, and Carlowrie continued to be an anchor and haven, but ultimately a millstone.

Isobel returned to writing, painting, and the lecture circuit. In the autumn of 1931 there came an invitation that struck terror into her generally fearless heart. She was invited to Cambridge to speak to a geography class taught by Professor Frank Debenham, director of the Scott Polar Research Institute, the most august body of northern explorers and scientists in Britain. On her return to Carlowrie she wrote: "[It was] not at all so formidable as I had dreaded! You were all much more kind than I merited and I am sorry I transgressed a whole half-hour longer than I should have done—a horrid sin in a lecturer. I can only apologise as I had broken my watch-glass at dinner and the watch had stopped."[5] The reply from Debenham was enthusiastic: "You held the audience enthralled, and you need not have been concerned about speaking too long or too fast. The reason for your vividness is that you are so obviously in love with Greenland and the Eskimo, and your descriptions ring true."[6]

At Cambridge Isobel was in the company of men of extraordinary polar experience. Frank Debenham was the geologist in the land exploration party of the Scott Antarctic Expedition on which the leader perished after reaching the South Pole in 1912. The Scott Polar Research Institute at Cambridge, founded as a living memorial to Robert Falcon Scott, was his inspiration. She was entertained in the home of J. M. Wordie, a fellow Scot who had not only been on Sir Ernest Shackleton's expedition to the Antarctic in 1914–17 but had recently led expeditions to eastern Greenland. Friendship with these men would continue, resulting in the gift of some of Isobel's Greenland artifacts to the museum of the Institute and a valuable connection with Cambridge.

About this time Isobel received a book that she would later describe as the greatest treasure in her library. It was Knud Rasmussen's *Across Arctic America,* inscribed to her by the author. Published in 1927, the book included Rasmussen's narrative of his sled journey in 1923–24 from Hudson Bay to the edge of Siberia with two Greenlanders, Miteq and his cousin Anarulunguaq, a young woman. Beginning at Repulse Bay, an inlet on Hudson Bay, they traveled the Arctic coast as far as Nome, Alaska, attempting also to visit Siberia, where permission to land was refused. They visited every Eskimo group on the route, collecting their stories and traditions, with a view to showing the kinship of the people from Greenland to Alaska. This was part of the Fifth Thule Expedition, a Danish mapping and ethnographic exploration begun in 1921. In contrast to the scientific report, *Across Arctic America* is Rasmussen's popular account of the expedition, written with humanity and without dwelling on the hardships of the trip.

With Rasmussen's book as her inspiration, Isobel began to study maps, read, and reflect on her next foray into the North. A letter to Frank Debenham gave the first hint of her intentions: "I am planning a new journey at last . . . and am hoping to visit the races of the Aleutian Isles off the Alaskan coast. If funds permit I might go on to Point Barrow and see something of Alaska, so if you would like any work done amongst the Aleuts do let me know."[7] She also mentioned that she was hoping to get a commission to collect seeds but was really going to study the people and to get a new subject to lecture on, having given the Greenland talk some sixty-five times. She was hoping to get to Attu Island, the farthest of the Aleutians, and would find out what was possible when she reached Vancouver or Seattle.

Despite her intention to leave early in May—only three months hence—Isobel was remarkably vague about her plans. The thought of visiting not only the westernmost island of the Aleutian chain but also Point Barrow, the most northerly point of mainland Alaska, and of being away only six months showed a complete lack of understanding of the transportation and conditions in the area she was considering. Certainly any ship going from Seattle to the Bering Sea had to pass through the Aleutian chain, often making a stop at Dutch Harbor, but to expect to reach the outer limits in two directions was highly optimistic. It also showed that she was prepared to set out without firm plans in hand and improvise as she went along.

Equally interesting is that she was offering to make ethnological observations based on her stay in Greenland, where she found the natives "a most interesting and delightful people." She accepted Rasmussen's theory of the racial link between all the Eskimo peoples and proposed her own theory that they came across from Asia using the islands of the Aleutian chain as stepping-stones. There is no record of Professor Debenham's reply, but he obviously took her seriously enough to report her intentions in the *Polar Record*.[8] In addition, the curator of the Museum of Archaeology and Anthropology at Cambridge University provided her with money to buy artifacts for the museum.

Soon after her letter to Debenham, Isobel visited the Royal Botanic Gardens at Kew to show the scientists samples of her work as a plant collector and to ascertain whether they would be interested in supporting her on a trip to northern Alaska. She explained that her first trip to Greenland had been as a private botanist and that her second trip was partly financed by several keen horticulturalists, the late Lords Dewar and Forteviot, and the Royal Horticultural Society, for whom she collected over fifty varieties of seeds.[9] The Kew administrators were interested in plants from northern Alaska and provided her with a collecting outfit consisting of two presses with straps, a supply of drying paper, thin laying-out paper to place between the specimens, and ten field notebooks, as well as a letter authorizing her as a plant collector for the Royal Botanic Gardens. They were prepared to pay her £2 10s. per hundred good specimens but could not afford more than a total of £15.[10]

After her contact with Kew there is no further mention of going to the Aleutians, and Isobel began to develop a clearer picture of the route she wished to take. But the transportation was not fully settled, and her mind seethed with possibilities. The plan she outlined to the director of Kew was to sail to Vancouver, go on to Whitehorse in the Yukon, "travel right up [*sic*] the Yukon from Whitehorse to St. Michael's near the Bering Strait, from there to Nome, and from Nome to Point Barrow, where I shall be *probably* [my italics] picked up by a boat of the Hudson Bay Co. expected there in summer. I shall travel down [*sic*] the Mackenzie River by the last boat in the autumn and so home, I expect, via Edmonton and Canada about November."[11] In a later letter to Kew she added that she had obtained a visa from the USSR and had hopes of penetrating into East Cape, Siberia, but would await the outcome of the show trials at Moscow before venturing into Soviet territory, although she assumed that the Siberian frontier would be quiet.[12]

Even with plans made and some of the transportation booked, Isobel was open to ideas for extending her travels farther into unknown territory.

During her stay in Greenland she had met an American, Reg Orcutt, who traveled the world on business and pursuing his own interest in photography. A member of the Explorers Club in New York and friend of the explorer Vilhjalmur Stefansson, Orcutt had already outlined a route for Isobel to follow, and she had booked her passage on the shipping line he suggested. When he arrived in Edinburgh in mid-April he was carrying a letter from Stefansson received as he boarded his transatlantic ship. Stefansson assumed that Isobel was planning to winter at Point Barrow (the first time this idea had been aired) and suggested that she would find it comparatively tedious, varying too little from what she had already seen in Greenland. He proposed that she would find it more interesting to winter near his friends the Masons on the Porcupine River below the tree line, where she would experience much colder temperatures and see more wildlife.[13] Isobel immediately sent a letter and a telegram across the Atlantic asking for more details, and her response to Stefansson's reply shows that she was in a lather of excitement:

> I think I can arrange to be away from home over the year . . . it would be most interesting to have a little cabin for myself built near the Masons as you suggest, and yet in touch of the Eskimos and Indians whom I would like to study. . . . I would rather have it than living with anyone else, yet perhaps it could be near enough to the Masons so that I could have one or two good meals a day with them, and so not to have to fill my time with too much cooking. Or else, perhaps I could have . . . an Indian or Eskimo woman to be my attendant for not too great a sum to do the cooking? Would it be near enough to an Indian settlement for this, if she went home at night? . . . If I had a really capable and honest Eskimo or Indian woman to come every day and be my attendant and cook that would be most interesting of all, and to have the Masons as near neighbours would be delightful. . . . I would like to meet the animals on friendly terms and do not shoot at all, so I hope there are not too many wolves about![14]

The letter was written less than two weeks before Isobel was due to sail, and she gave the shipping line as her return address.

Stefansson was very understanding in his reply, recognizing immediately that Isobel's ideas were based entirely on her Greenland experience. He warned her of the unpredictable nature of transportation in the Far North, unlike the west coast of Greenland, where it was possible to keep to a schedule. He also explained that, unlike Greenland, where a colonial government had accustomed the native people to European ways, it was almost unthinkable to hire a woman to come regularly to cook and clean. He did, however, continue to suggest that Isobel try to reach Aklavik by going up the Porcupine River and making the eighty-mile portage down the Rat River in winter, even though she had told him she would have three small trunks in her luggage. He urged that, whatever she decided, she should try to consult the Masons when she reached Fort Yukon on her way down the Yukon River.[15]

Always open to suggestions, Isobel was prepared to give this new idea consideration and make up her mind at the appropriate moment. On paper, her original plan was elegant in its simplicity: boat travel all the way and all booked in advance, except for the little piece between Nome and Point Barrow. How differently it turned out!

The trip began in a leisurely fashion. On a 9,500-ton cargo boat, the *Pacific Shipper*, Isobel sailed from Manchester on May 3, 1933, down the ship canal, across the Atlantic by the southern route, through the Panama Canal, and north along the west coast of America. The last of the other six passengers left the ship at Victoria, leaving her to carry on to Vancouver, where she arrived five weeks out from Manchester. It had been an idyll of star gazing, watching the sea life, even some hurried sightseeing at the ports of Los Angeles and San Francisco. Approaching Vancouver, a realization of the unknowns ahead surfaced as the pilot of the ship asked about her plans. Hearing that she was setting out alone for Point Barrow, he expressed surprise and admiration at her bravery.

When she explained that the Hudson's Bay Company had agreed to pick her up at Point Barrow on or about July 22 in the supply ship *Anyox*, the pilot remarked that at the moment it was in port being fitted out. He also mentioned that the Hudson's Bay Company had lost two good boats in the area of Point Barrow in the past three years, the *Lady Kindersley* and the *Baychimo*.

Engine trouble had delayed their arrival in Vancouver by three days, leaving Isobel only two days to make all the arrangements for the journey ahead. At the Hudson's Bay Company store she ordered canned goods and supplies to take her through the Arctic winter in case she was stranded there, and these would be delivered directly to the *Anyox* when it sailed from Vancouver.

The boat trip up the Inside Passage to Skagway was easily arranged, as well as rail and boat transportation to Dawson in the Yukon. Beyond Dawson arrangements were hazy, river transport on the lower Yukon River being unscheduled. She was allowed to book only as far as Tanana in Alaska, since tourists did not usually go beyond that point. The itinerary north of 60°, chosen with the aid of the atlas at Carlowrie, now had some significant gaps and more than a few question marks, which would require decisions and flexibility.

When the captain of the *Pacific Shipper* came to the Hotel Vancouver to say good-bye to Isobel, he noticed that an international scientific gathering—the Pacific Congress—was in progress, with lectures and meetings in the large salon off the hotel lobby. Thinking that as a botanist she would be at home in this gathering, he sought out the manager of the hotel, who in turn introduced her to Dr. Hutchinson, professor of botany from the University of British Columbia. She was immediately taken off to be introduced to other North American botanists and entertained at a luncheon with people of common interests.

When it was found that she was heading for the Arctic coast, she was introduced to the distinguished anthropologist Diamond Jenness, who had spent several years living among the Eskimo in that area. He explained that she must obtain a scientific exploration license from the Canadian government because she intended to collect plant specimens for Kew and ethnographic artifacts for Cambridge University.

Isobel complied by writing to the Department of the Interior while on board the *Princess Norah* on the eve of her departure for the Yukon. Already her ideas had strayed from the plan she had outlined to the director at Kew: "I hope to spend the winter on Herschel Island or at Aklavik (Mackenzie Delta). I am collecting flora for Kew and am also making an ethnographical collection of Eskimo specimens for Cambridge Museum England. This morning at the Scientific Congress here, I met Dr. Diamond Jenness who suggested I should apply to you for permission to dig in the Eskimo ruins in the Arctic near Herschel or Mackenzie if necessary on behalf of Dr. Clarke at Cambridge."[16]

This is yet another example of Isobel's susceptibility to suggestion. Her visit to Greenland was prompted by suggestion; now, after a conversation with Diamond Jenness, she was asking permission to winter in the Arctic and to dig for artifacts. Jenness was sufficiently impressed by her that when later contacted at the National Museum in Ottawa he recommended she be granted a permit under the Eskimo Ruins Ordinance, with the proviso that no skeletal material could be collected.[17] In Canadian scientific circles in 1933, the magic words Kew and Cambridge carried the weight necessary to open any doors. Late in August it was noted that Dr. Jenness had heard nothing from Miss Hutchison and was of the opinion that she had not carried out her proposed plan.[18]

Ironically, Isobel received permit 22, dated June 30, 1933, months later in Aklavik, along with a letter advising her that the matter of accommodations and means of travel was a personal one and that she would need to obtain food, clothing, and travel requirements in advance.[19] By the time she received this message she had become an old hand at Arctic travel.

The *Princess Norah* was a luxury boat, crowded with tourists and those going north to work, including Harry Lester, in the scarlet-and-blue uniform of the Royal Canadian Mounted Police, and the Mother General and three Sisters of Mercy, all traveling to Dawson. The boat made brief stops at Alert Bay at the north end of Vancouver Island, at Prince Rupert on the British Columbia mainland, and at Ketchikan in Alaska, where Isobel hurried off to collect plants in the pouring rain.

The boat continued on to Skagway, where Isobel boarded a train. The seven-hour rail journey through White Pass followed the route taken by many of the gold seekers in the Klondike stampede of 1898. Most of the passengers got off at Carcross, leaving Isobel in the parlor car with Angus Whitehouse, a poet-entomologist, the young Mountie, and a few old-timers. The talk naturally turned to travel. Isobel mentioned Stefansson's proposed alternative route up the Porcupine River, wintering at Old Crow, an Indian village near the tree line, and crossing the continental divide to the Mackenzie River. One of the old-timers advised her not to miss seeing Nome and the Bering Sea. He recommended that she try to sail with Captain T. C. Pedersen, who made the trip from San Francisco to Herschel Island via Nome each year, but said she would need to reach Nome early to catch him.

Once again suggestibility was at work in Isobel. Every idea had possibilities. All would be considered and the choice made at the last moment.

At Whitehorse she boarded the wood-burning paddle wheeler *Casca* for the comfortable trip down the Yukon to Dawson. The frequent stops to take on fuel at the woodpiles along the way provided opportunities to go ashore and add to her plant collection. The leisurely trip gave her more time to discuss travel options, with opinions equally divided between the advantages and disadvantages of wintering at Old Crow or Herschel Island on the Arctic coast. Wintering had become a definite expectation.

Arriving in Dawson, Isobel found that the American riverboat *Yukon* was delayed downriver, and she would have a week in the "queer town of broken old wood shacks,"[20] greatly shrunk since the heyday of the gold rush. In 1933 gold mining was still the predominant industry, though employing fewer miners and controlled by one large company using superior technology. During the 1930s a permit system was established to regulate native people's access to Dawson, where they were welcome only to sell the products of their hunting, fishing, and trapping and to visit government officials. As the territorial capital, Dawson had a small core of year-round residents and an influx of transients each summer with the opening of navigation on the Yukon River.[21]

Isobel was the only guest at the Royal Alexandra Hotel, and the manager introduced her to many people of interest. A few days after she arrived, summer erupted with a vengeance; the temperature reached 100°F in the shade and made walking any distance a penance. Before the heat wave she climbed the Dome, 1,500 feet above Dawson, in the company of Harry Lester, the Mountie she had met on the boat, and encountered her first and only black bear. She met other botanists, Mrs. Hartshorn and Mr. Delagrave, the town shoemaker, saw their collections, and collected specimens of her own. Colonel Allard, superintendent of the RCMP at Dawson, and his wife, Dorothy, entertained her several times in their home and drove her in their car to see the gold workings on the Klondike River.

In spite of the vastness of the North, news vital to northerners is disseminated without apparent effort. The *Anyox* was to be the Hudson's Bay Company's only supply ship to the western Arctic that year, and Isobel had already heard speculation about its seaworthiness from fellow travelers on the way north. It was a new ship with a wooden hull, built to carry copper concentrates on the Pacific between Chile and Canada, but it had been chartered by the Hudson's Bay Company after the boat chartered in 1932 to replace the *Baychimo* had proved unsatisfactory.[22] The *Anyox* was about to be severely tested on its maiden voyage. Colonel Allard advised Isobel not to trust her winter provisions to the *Anyox* but to have them shipped by the direct route down the Mackenzie River to Aklavik, and he arranged to make this change for her by a radio message to Vancouver.

At last, on June 23, the *Yukon* arrived, and Isobel began the next leg of her downriver trip. At the American border, at the village of Eagle, well after midnight, the customs and immigration official tapped on the mosquito netting of her cabin, asking her destination:

> "Herschel Island by Nome and Barrow," was her reply.
> "And how do you expect to get round from Nome?"
> "By the Hudson's Bay Company boat *Anyox*," I answer glibly. (Though the *Anyox* only calls in at Barrow, and my journey thither from Nome is as yet a missing link, of what use to quibble over such trifles at one o'clock in the morning?)[23]

Arriving at Fort Yukon at midnight, Isobel met Stefansson's friends, Willoughby (Bill) Mason and his wife,[24] who spent their winters trapping on the upper Porcupine River not far from Old Crow. At this point she was only three hundred miles from Aklavik, which could be reached by charter airplane or by canoe with the expensive and difficult portage across the divide into the Mackenzie valley. During the one-hour stop the Masons gave her a tour of the village, and after discussing various possibilities Isobel decided that the journey via the Porcupine River did not hold the same prospects for adventure as the long detour down the Yukon, through the Bering Strait, and around the Alaskan coast. One more decision had been made for the route ahead, but more would be needed almost immediately.

Before they reached Tanana it became known that the new boat to take her to the Alaskan coast was running late. She was advised to stay on the *Yukon* to Nenana, where there were better accommodations, and either wait for the riverboat or take the train to Fairbanks and fly to Nome.

Her decision was to let fate decide. If the downriver boat was at Tanana, she would transfer to it; otherwise she would go on to Fairbanks. Isobel was a strong believer in Providence. Being prepared to accept cheerfully whatever was provided became the key to the success of this enterprise.

Stefansson, realizing the difficulties she might have in carrying out her original plan, had already been in touch with Alaskan Airways at Fairbanks on her behalf. A plane was leaving for Nome on the very afternoon she

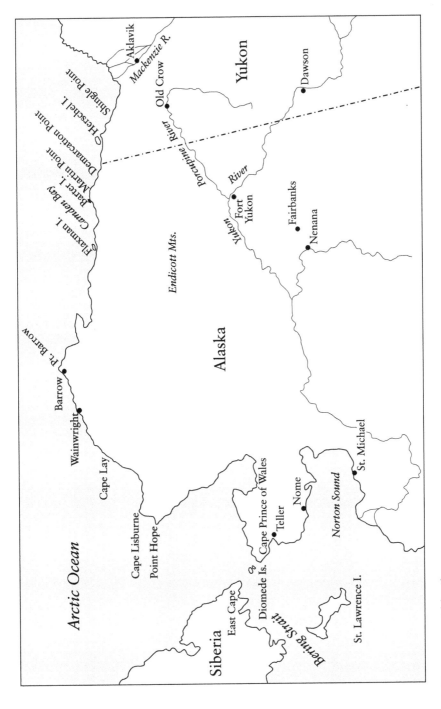

Dawson to Aklavik

arrived in Fairbanks. The price of the ticket, plus an equal amount for transporting her three hundred pounds of baggage, gave her a moment's hesitation. Her impression of the town was that it had been wrongly named: "Her banks are anything but fair, as I found to my cost presently when I tried to find the fluctuating rate of the British pound sterling as against the Fairbanks dollar, and dropped about a hundred dollars in the process into Uncle Sam's very tightly-buttoned pocket!"[25]

With one previous fifteen-minute experience of flying, Isobel settled back in the four-passenger plane to enjoy seeing the Yukon valley from the air.[26] As they approached the coastal mountains the weather closed in, and they were forced to land at Nulato to wait for news of clearing over the Bering Sea. The following day the clouds over the six-thousand-foot mountains parted long enough to allow the flight to proceed, and they landed at the mouth of the Nome River.

Facing Norton Sound, on the south side of the Seward Peninsula, Nome is closer to East Cape in Siberia than to most communities in Alaska. Nome, like Dawson, was created by the race for gold. Before the turn of the century, a few Klondike-bound prospectors, stopped by freeze-up at the mouth of the Yukon River, searched the area of Norton Sound and found enough gold to make it worthwhile to register claims. The secret was soon out, and by the winter of 1898–99 there was a tent town housing 250 prospectors mining inland creeks. Then it was discovered that the black sands of the beach fronting the ocean carried as much gold as the creeks and that digging was easier. In the summer of 1899, 2,000 miners working along a forty-two-mile stretch of beach took out $2 million worth of gold.[27] Unlike the Klondike, where the Canadian government sent in a force of nearly a hundred Mounties at the height of the gold rush, the "city" of Nome was the latest frontier boomtown and "as lawless a community as ever existed in the West."[28]

By the summer of 1933 the excitement on the beach had subsided and Nome, like Dawson, had shrunk and stabilized, with a cosmopolitan white population of about 900 in winter. The town presented a dilapidated appearance as Isobel arrived by taxi at the Golden Gate Hotel, a relic of the days when Nome had a population of 40,000. Circumspect in her descriptions and never one to complain, she only mentioned that it was "a large flat-fronted building that stood slightly askew." Richard Bonnycastle of the

Hudson's Bay Company, reaching Nome from the ill-fated *Baychimo,* was more outspoken: "Dilapidated, old-fashioned and poor . . . the wallpaper is hanging from the walls, the doors are awry, and the carpets are faded."[29] Dr. Aleš Hrdlička, on an ethnological expedition for the Smithsonian Institution, gave another picture: "A big old frame building, so badly out of plumb in several directions that one almost hesitates to walk by it. . . . [My room] smells musty, of old sweat, old blankets, and mold, and looks out on a dilapidated tin roof."[30] Although she did not know it when she arrived, this would be Isobel's home for the next five weeks.

The gateway to the Bering Strait, Nome was the first community of any size to be reached by explorers on their way to the Arctic or returning from there. Roald Amundsen, in the *Gjoa,* arrived in 1906 after the first successful negotiation of the Northwest Passage. Ejnar Mikkelsen was feted by the town in 1908 after a heroic journey by dog team around the north coast, where he had made an unsuccessful search for new land out on the ice of the Beaufort Sea. Vilhjalmur Stefansson came in 1912, and Knud Rasmussen in 1924 at the end of his journey across Arctic America. Rear Admiral Richard Byrd passed through on the way to his flight over the North Pole in 1925. All of these people not only passed through Nome but were entertained by the hospitable Lomen family.[31]

Carl Lomen and his father, a lawyer from Minnesota, had gone to Nome in 1900 for a summer of prospecting. They were captivated by the North, and after three years they sent for Mrs. Lomen, Carl's four brothers, and his sister. Nome needed an honest lawyer, and G. J. Lomen fit the description. Eventually he was appointed to the bench and served the area as a judge. The Lomen sons operated as a cooperative, first buying a photographic studio and next a drugstore. Their most far-reaching enterprise was in domesticated reindeer; they acquired their first herd in 1913, and by the 1930s they owned roughly a million animals. Later in her travels Isobel would meet people involved in transferring part of that herd to a Canadian government reindeer station near Aklavik.

Within hours of arriving in Nome, Isobel met many of the town's leading citizens—the Lomens, Ruth Reat, who became a lifelong correspondent, and others—finding kindness on all sides. She appreciated the assistance of Harry Lomen at the photographic studio after days of fieldwork, collecting and photographing on the barren slopes above Nome. She met a woman flier, Edna Christoffersen, and because Isobel had flown to Nome she herself was made an honorary member of the New Women's Aviation Society, which she labeled an example of American hustle.[32]

Nome was a friendly, hospitable place, but it was a "white man's town in Eskimo territory."[33] The Snake River separated the Eskimo village from Nome by both geography and spirit. Knud Rasmussen, who enjoyed the hospitality of Nome, discovered the latent discrimination when he was refused service at a restaurant at the end of his journey because his two Greenlandic companions were with him. "It is always thus on the frontier where the white man comes into contact with those whom he is pleased to term an inferior people. White women are generally the indirect cause of it. When they arrive on the scene the social lines become firmly drawn between those who associate with natives and those who do not."[34] This would not be the case farther north on the Arctic coast beyond the frontier.

Like the hotel in Dawson, the Golden Gate Hotel did not serve meals, and Isobel patronized the three restaurants on the main street, where she met the local people, who were free with information and advice for her onward journey north. Almost immediately she learned that the winter had been the longest in years, that the ice was still thick as far south as Teller, in the first bay north of Nome, and that six walrus had been caught recently on the ice not far from Nome.

Isobel's dilemma is clearly stated in her letter to Dr. Cotton at Kew, dated July 13:

> I do not yet know if I shall be able to connect with [the] Hudson Bay Co. boat at Point Barrow as transport to Barrow here is hard to get, but I hope to do so, and if so may winter either on one of the Arctic islands (Herschel possibly) or at Aklavik in the Mackenzie delta, and shall endeavour to get as many more specimens as possible. . . . I *hope* to make connection and be somewhere there over Christmas and later. I had to *fly* from Fairbanks with all my baggage to get here in time for the boat and am now held up for lack of further transport and may have to try airplane again but it is not a good trip and the planes don't go much up into the Arctic, except in winter with mail.[35]

The director of Kew replied on September 16, in a letter addressed to Isobel Hutchison, c/o Royal Canadian Mounted Police, Herschel Island or Aklavik, to the effect that the specimens she had sent had arrived safely and in excellent condition. By that date Isobel's whereabouts were unknown to almost everybody but herself, but she too was in excellent condition.[36]

Isobel had arrived in Nome at the end of June, intent on finding a way to reach Point Barrow to keep her appointment with the *Anyox* on July 22, but the *Anyox* had refused to pick her up in Nome because of the lack of harbor facilities. Her hopes were now pinned on Captain T. C. Pedersen of the *Patterson*. His reputation as one of the best ice pilots in those waters had been earned over more than twenty years as an independent trader and one of the chief rivals to the Hudson's Bay Company on the Arctic coast.

Captain Pedersen was still in Dutch Harbor in the Aleutians when Isobel sent him a wire requesting passage to Barrow. His reply telling her that the *Patterson* carried no passenger license and had no accommodations was a severe disappointment. There were a few other ships that regularly sailed north each summer. Toward mid-July the United States revenue cutter *Northland* would be making its annual coastal inspection as far as Demarcation Point, the border with Canada, but it was already overcrowded with passengers, and arrangements should have been made well in advance through the proper channels in Washington. The hospital-supply ship *North Star* would be going to Barrow sometime in August, but Isobel was still working to the schedule imposed by the *Anyox*. This left the *Trader*, a small vessel belonging to a Nome merchant, Ira Rank, which traveled north each summer to trade at all the Eskimo villages along the way. Although her new friends advised that the *Trader* would be very uncomfortable, Isobel saw it as an opportunity for collecting both plants and artifacts.

When Isobel inquired at the Ranks' shop, which sold everything from fur parkas to chewing gum, she learned that the *Trader* was on a trading expedition to St. Lawrence Island in the Bering Sea. Mrs. Rank assured her that her husband would be able to accommodate her for his regular trip to the north coast. When he returned, Ira Rank not only agreed but promised Isobel that if they did not tie up beside the *Anyox* at Barrow he would take her to Herschel Island himself if the ice conditions permitted.

It had been a worrying week trying to arrange onward transportation. With everything settled by the middle of July, Isobel began to enjoy Nome. Her new friends entertained her at dinners, teas, and picnics. She met Charlie Brower, a prominent trader in from Barrow, a place she hoped to reach before long. To augment her mother's pony coat, which she expected to use in cold weather, she bought a drugget parka, mukluks, and mitts from the Lomens.[37]

There was time now for the pleasures of botanizing on the tundra. Behind Nome the tundra sloped up to Anvil Mountain, a thousand feet high, from which Isobel could see the snowcapped Bendeleben Mountains, the spine of the Seward Peninsula. When she was not looking up, she was dazzled by the carpet of summer bloom at her feet. In her search for specimens for Kew, she had a knowledgeable companion in Charles Thornton, commissioner of mines and an outstanding amateur botanist. With his guidance she collected more than 200 of the 278 known species of plants from the area, and she enlisted Thornton to send directly to Kew any specimens not found during the time she was there.

In addition to collecting plants, Isobel also had an opportunity to buy some valuable artifacts for Cambridge University, which she refers to as "curios." The population of Nome swelled to about 2,000 in the summer as Eskimo from all parts of northern Alaska assembled to trade during the tourist season. Among them were the population of King Island, a tiny dot of land in the Bering Strait whose inhabitants were famous for their fine ivory work. During the summer's trading season they lived along the beach to the east of the town in tents or under upturned umiaks. In 1924 Rasmussen found the King Islanders among the most distinctive of the groups he met. Hardy, frugal, industrious, obstinate, and independent, they clung to their ancient festivals and beliefs in spite of being converted to Catholicism.[38] Among the items Isobel purchased was a pottery bowl reputedly from a grave, which the old Eskimo woman selling it told her she had seen her grandmother make from blood, feathers, sand, and hair. It smelt unpleasantly of blubber, and Isobel thought it was a stone lamp, of which she had several. When she reached Barrow, however, she discovered that this was a rare and valuable acquisition. She was also present at a drum dance ceremony performed by King Islanders, where some of the oldest women bore blue tattooed lines from mouth to chin indicating their married status.

Before the end of the month the news reached Barrow that on July 26 the *Anyox* had struck heavy ice in the Bering Strait, seriously damaging its wooden bow.[39] It was rescued in the nick of time by the revenue cutter *Northland,* answering its distress signal from the port of Shishmaref on the north side of the Seward Peninsula. The *Anyox* was patched sufficiently to return to Vancouver under its own steam, carrying all the Hudson's Bay Company supplies for the eastern Arctic with it. Because of Colonel Allard's premonition about the *Anyox* in Dawson, Isobel's winter supplies were not on board, having gone to Aklavik by a different route.

After "a wonderfully interesting" five weeks at Nome, the ice conditions were judged ready for the *Trader* to leave for the north coast, and Isobel's real adventure was about to begin. In Stefansson's opinion, an adventure was a sign of incompetence. Isobel disagreed: "Not one of the things that happened to me between Nome and Herschel Island were things which I expected to happen, for the man who confides his course to Providence is always sure of adventure—always sure to find that Heaven's ways are infinitely more interesting and adventurous than his own plans!"[40]

It was three months since Isobel sailed down the ship canal at Manchester, half a world away in distance and even farther in culture. In moving from Dawson to Nome, she observed the difference between a well-regulated region under the control of the Canadian Mounted Police and the wide-open frontier mentality of the Americans. This difference would resurface noticeably later in her journey when she returned to the domain of the Mounties. The stay in Nome was a useful transition from the world where transportation kept a schedule to the Arctic, where ice is king and ships enter only with its permission, obeying its laws.

Most of the travel to Nome was standard fare, in the company of regular travelers and tourists. When she came to write about her adventure in *North to the Rime-Ringed Sun,* she used the outdated convention of identifying the six passengers by occupation or a descriptive adjective, capitalized, labeling herself "The Botanist" on the first leg of her journey on the *Pacific Shipper.* She continued this technique, almost without exception, until she reached Dawson, where people began to take on names and personalities. It is possible that the disparaging reviewer writing in the *Saturday Review of Literature* did not read beyond Isobel's arrival at Dawson, for he found her book "consistently drab" and filled with "mousey details"—the only reviewer to treat it so harshly.[41]

Only after Dawson, at Tanana, did Isobel Hutchison begin to show the tenacity with which she would hold to her course once her decisions were made. At Nome she demonstrated flexibility and willingness to try anything to achieve her goal, and as in Greenland, she was open to every new experience. After Nome the real adventure was to begin.

8

Into the Ice

At last Ira Rank judged that the ice conditions in the Bering Strait would permit the *Trader* to begin its 1933 trading trip north. The distance from Nome to Barrow is just over five hundred miles, and with stops at small Eskimo settlements along the way, Rank predicted that with uncommon luck it would take only about five days.

The *Trader* was a ten-ton schooner, seventy feet long, powered by a forty-horsepower motor and fitted with sails. Isobel described it only in terms of its limited accommodations. A passenger of the previous summer gave a more vivid description: "a dirty little schooner, reeking with the smell of oil and seal, covered with sea-growth, the name just visible through a scab of rust on the bow."[1] David Irwin, a young man vagabonding across the North in 1932, was given passage on the *Trader* from Nome and put ashore at his own request on the Arctic coast east of Barrow to make contact with the reindeer drive. Isobel had heard much about the reindeer drive from Carl Lomen in Nome and hoped to see the huge herd somewhere on her travels.

Ira Rank, the vessel's owner and trader, was a tremendous figure of a man. His wide, sloping shoulders tapered into thick arms and large hands; his bald head was tanned a deep brown, and his friendly light blue eyes met one with a direct gaze.[2] Born in Russia, he had escaped from the Kamchatkan coast of Siberia to Nome when the Soviet revolutionary forces occupied the trading post he was operating with the co-owner of the *Trader*, a Russian-born naturalized American. Rank now traded regularly on the islands in the Bering Strait, but he was careful not to cross the narrow passage between the American island of Little Diomede and the Russian Great Diomede, since there was reputed to be a price on his head.

Ira Rank concentrated on trading and left the running of the *Trader* to the two Palsson brothers from Iceland. Kari, the younger, was engineer and chief cook, while Pete was the captain. Pete Palsson, good-natured and capable, had also been a trader on the Siberian coast, trading as far north as Wrangel Island. His boat had been confiscated at the time of the revolution, and he was lucky to escape with his life.[3] One of Palsson's partners in this enterprise was Gus Masik, who would provide an important link in Isobel's journey farther north.

The boat was tightly packed with all manner of trade goods and canned delicacies, as well as sweet corn, lobster, bully beef, boxes of fresh apples, oranges, grapefruit, and even watermelon. The Eskimo customers had an increasing appetite for such things, as well as for radios, concertinas, gramophones, and silk stockings. The hold was filled with tobacco, which would all be sold by the end of the voyage. Even if they were caught in the ice pack, they would eat well with so much food on board.

Accommodations on the *Trader* were snug. The cabin served as the galley and as the sleeping quarters for Ira Rank and his passenger. They slept in narrow bunks on either side of the table, and the mainmast descended through the deck into the center of the table. The deck was so close above the bunks that sitting up in bed was impossible. The Palsson brothers slept close to their work: Pete had a bunk above the lockers in his glass-walled pilot house on deck, and Kari slept beside the engines at the stern of the boat. Isobel had come a long way from the days when sharing a bedroom even with her sister was unthinkable.[4] Whereas she used to complain in her diary about having to share a cabin at sea, Isobel's experience in Greenland had taught her to accept circumstances as she found them. The anxious time in Nome finding a boat that would take her north made her appreciate the *Trader* regardless of the accommodations. On meeting Ira Rank and his wife in Nome, Isobel recognized him as someone she could trust, and since he was accustomed to sharing accommodations with a passenger there was no awkwardness in the situation.

Sailing into the Bering Strait on August 2, the *Trader* encountered the ice pack almost immediately, and on their first night out of Nome a gale forced them to run for shelter in the bay behind Cape Prince of Wales. They had not progressed beyond the Seward Peninsula, and the winds held them there for two days, a harbinger of things to come.

The days at anchor gave Isobel an opportunity to collect plants on the hillside above the beach, and the time ashore was a welcome escape from the storm-tossed cabin. Two miners, Maclean and Waldhelm, operating what they claimed was America's only tin mine nearby, had a comfortable shack on the beach, and like most northerners they were glad of company, entertaining the four from the *Trader* to supper both evenings and regaling them with stories.

Cape Prince of Wales, only fifty miles from East Cape in Siberia, was the point of departure for travelers trying to cross the Bering Strait. One of Waldhelm's stories was about a woman, traveling alone on foot, who was attempting to cross the strait to reach her home in Europe. She had come down the Yukon from Fairbanks, where she had been working as a waitress, and Ira Rank remembered bringing her on the *Trader* from the mouth of the Yukon River as far as St. Michael on Norton Sound. Her clothes were in rags by the time she reached Cape Prince of Wales. She was thought to be Dutch; the tin miners had found some of her clothes and a letter addressed to someone in Holland on the hillside, but there was no trace of the woman herself.

The story may be that of Lillian Alling, a reclusive woman suffering from homesickness in New York City in 1926, who set out to walk home via the Yukon and Bering Strait. She reached Dawson in October 1928, wintered there working as a waitress, bought an old skiff, and went down the Yukon River as soon as the ice was out in the spring. An Eskimo sighted her plodding northward, pulling her two-wheeled cart, near Teller on the coast north of Nome. Lillian Alling was reputed to be Russian, but that nationality may have been inferred because of her effort to reach the Siberian coast. Her fate has remained a mystery, although there was a report of Soviet officials' questioning a woman in European or American dress with three Eskimo men from the Diomede Islands in the fall of 1930. The woman said she had come from America, where she had not been able to earn a living or make friends, and she had walked most of the way. The Eskimos had brought her across the Bering Strait from Alaska. It is tempting to believe that the Dutch woman and Lillian Alling were the same and that she completed her incredible journey.[5] In any event, this story of the lonely, determined marathon walker struck a sympathetic chord in Isobel.

After being stopped for two days, the *Trader* rounded Cape Prince of Wales with a fair wind and continued north through mists that hid the Diomede Islands. "On this day for the first time we seemed to smell ice. The sun shone through the haze like a drawing in silver-point. Puffins flew in small companies over the bleak waters; sometimes an Arctic tern, most graceful of sea-fowl skimmed above us. Through the mist, towards evening, gleamed a pale-grey rainbow."[6] After a calm day they reached the next prominent point of land, Point Hope, just as the sea began to toss them about.

Point Hope is one of the most interesting Eskimo villages in Alaska, built on the ruins of several older civilizations of undetermined age. Knud Rasmussen, traveling in the opposite direction along that coast by dog team nine years earlier, had found a great store of folklore in this village. At one time more than 2,000 people had lived in the area, supported by whaling.

Whales, seals, and walrus were the mainstay of the Eskimo of the north Alaskan coast, and marine mammals had supported a large and vigorous population. Beginning in the middle of the nineteenth century, the first American whaling ship entered the Bering Strait, and within four years there were more than 220 ships in the area. This incursion had a disastrous effect not only on the whale stocks but on the lives of the Eskimo.

As whales became harder to find in the late 1860s, the whaling ships began an intensive harvest of walrus, both for their oil-producing blubber and for their ivory tusks. Walrus were the staple commodity of the Eskimo who lived on the islands of the Bering Sea, and they used every part of the animal for their survival. When it was discovered that 1,000 people out of a population of 1,500 had starved on St. Lawrence Island during the winter of 1878–79, there was a gradual move to ban the walrus hunt.[7]

The contact with the men from the whaling ships had an equally disastrous effect on the Eskimo population of the coast. Although it is suggested that the Eskimo already knew about alcohol from their traditional trade with Siberia, their exposure to it now escalated, and they also learned the art of distilling it from molasses and flour. Ugly incidents occurred at places like Cape Prince of Wales, where drunkenness during trading had resulted in Eskimo deaths and a vendetta against white men that lasted for years.[8]

In the beginning, the whaling ships sailed out of New Bedford in New England; later they would use the port of San Francisco. The earlier voyages, around Cape Horn and north through the Pacific, could last up to four years, in contrast to those of British whalers in Greenland waters, which were measured in months. The method of hiring the American crews was to have "landsharks" pick up men from the lowlife that drifted around seaports and keep them inebriated until the voyage was under way. Too late the reluctant sailors discovered their fate. Men who jumped ship in the South Seas were replaced by natives of the Azores, Fiji, and the Caroline and the Sandwich Islands. Unruly sailors of a motley background, controlled by brutality, were the first outside contacts for many of the Eskimo of the north Alaskan coast.[9]

What protection did the American government give the natives during this time? The purchase of Alaska in 1867 had been engineered by Secretary of State William Seward over the objections of the House of Representatives. There had been no public demand to purchase Alaska, and legislators were reluctant to spend tax dollars on what they considered an unknown and useless country.[10]

It was not until 1879 that the first annual patrol was made by the Revenue Cutter *Corwin* in an attempt to bring a semblance of law enforcement to the area. It has even been suggested that some of the early revenue cutters carried on the liquor trade with the natives.[11] The revenue cutter was still the chief symbol of government presence along the coast, and it was making its annual tour at the same time as Isobel was traveling north with the *Trader*.

While Ira Rank was trading in Point Hope, Isobel and Mrs. W——, the wife of the schoolteacher, visited the archaeological excavations and collected plants on the flat tundra meadow. Until recently, human bones had lain scattered over the surface, but most had been collected and interred, although bones could still be found in the grass. Isobel purchased artifacts for Cambridge, and the time was all too short for the interesting material the area contained. Their quick departure was governed by the wind and the ice, though Isobel would have liked to spend several days in Point Hope.

Without stopping, the *Trader* passed Cape Lisburne, an imposing sheer headland rising nine hundred feet from the water, and headed across a wide stretch of open water to Cape Lay. Cape Lisburne was a dreaded barrier to those traveling around the coast. Rasmussen had rounded the cape with some Eskimo in a skin boat so rotten that it had holes stuffed with scraps of reindeer skin and woolen comforters.[12] In the winter of 1907 Ejnar Mikkelsen was nearly swept away by the sea as he attempted to edge his way around the base of the cliffs with his dogs.[13]

"As we neared Point Lay we set eyes on our first ice-cake—met at this early date in August very much farther south than usual," Isobel wrote.[14] She had the happy faculty of trusting those in control: having committed herself to travel on the *Trader,* she gave herself up to enjoying the experience, leaving the problems of sailing to the crew. With the eye of an artist and the soul of a poet, she was completely absorbed by the beauty that surrounded her.

The Arctic coast of Alaska presents a difficult piece of navigation. In addition to the shifting ice pack, a long line of shoals hedge the coast from Cape Lisburne to Point Barrow. Within these shoals a chain of shallow lagoons made it possible for a small vessel like the *Trader,* drawing only five feet of water, to shelter when the wind drove the ice pack inshore. Many of the shoals were uncharted, and Pete often had to watch from the mainmast, shouting directions to Kari in the wheelhouse below, while Ira took soundings.

They passed Cape Lay with the ice barrier thickening fast, the west wind driving it inshore, and anchored for the night to the largest ice cake they could find. By morning the ice had carried them twelve miles back the way they had come. With the gale raging, they sought shelter behind a sandspit in the company of the *Holmes,* a ship that depended entirely on sail, carrying supplies from Vancouver to Barrow. The revenue cutter *Northland,* on its inspection tour of the Alaskan Arctic coast, passed outside the shoal. The ships going north would be continuously in contact, ahead of or behind each other at the whim of ice and wind. With complete faith in the crew, Isobel was exhilarated by the natural forces in play around her and interested in watching all that was happening.

By August 11—nine days out from Nome—the *Trader* arrived at Wainwright, one of the largest Eskimo settlements on the Alaskan Arctic coast. They had no sooner cast their anchor in the bay, after a rough passage, than Dick Hall, a local trader, came out in a skiff with a couple of natives to tell them that the ghost ship *Baychimo* had been sighted only twelve miles offshore. Hall and the two Eskimo came aboard the *Trader,* the anchor was lifted, and the little boat followed a lead out through the ice floes. "It was the thrill of a lifetime, and I watched spellbound the myriad colours of emerald green upon the fantastic ice-cakes, lovelier than any water lily. *Trader* halted in her slow progress at one giant floe to replenish her water-barrel from a sapphire pool in its heart."[15]

The ghost ship was a mere smudge on the horizon, twelve miles away. The intervening sea was thick with drift ice heaving in the swell, barely allowing the tiny *Trader* a passage through it. The *Trader,* which until now had crept along the coast close to shore, was heading out into the ice-choked ocean. Had Isobel been a fainthearted person, she could have asked to be put ashore while the *Trader* pushed its way through the ice floes to look for the *Baychimo.* But she had long since gained her sea legs, and seasickness was not a problem. In the nine days she had been aboard the trading vessel, she had been accepted as a member of the crew, and the question of putting her ashore was not even asked. Isobel was as keen to be part of the adventure as they were.

The *Baychimo* was a large steel-hulled oceangoing ship with an interesting history. Built in Sweden in 1914, it was ceded to the British government by Germany as part of war reparations and bought by the Hudson's Bay Company in 1921. With a crew of about thirty and space for ten passengers, it became the company's Arctic supply ship, first in the eastern Arctic, going to Pond Inlet at the north end of Baffin Island, and then serving the Siberian posts on Kamchatka. In the course of its career the ship circumnavigated the globe, using both the Suez and Panama Canals. By 1925 it was used exclusively to service the posts in the western Arctic.[16]

The *Baychimo* was returning south from its annual trip to Herschel Island in 1931 when it was caught in the ice sixty miles east of Point Barrow. It was a particularly bad year for ice, and the *Baychimo* had not reached Herschel Island from Vancouver until late in August. After three days of battling heavy ice in an attempt to call at a few of the company's posts east of the Mackenzie delta, Captain Cornwell, with eight years' Arctic experience, wanted to abandon the attempt, return to Herschel Island, and land the supplies for the eastern Arctic. The fur trade commissioner of the Hudson's Bay Company, Ralph Parsons, on board for that part of the trip, urged the captain to make one more attempt to get through—"egged him on without taking any responsibility," according to Richard Bonnycastle, the supervisor of the region.[17] As a result, the *Baychimo* did not leave Herschel Island for home until September 13, much later than Captain Cornwell felt was prudent. The ship fought its way around Point Barrow and was finally stopped by grounded ice inside a shoal off the Seahorse Islands, twenty miles from Wainwright, on September 24. Ice formed around the ship before it could move again, and it was firmly caught within walking distance of the shore. When it became apparent that the ship was frozen in for the winter, a cabin fifteen feet by forty-five was built on shore, using lumber and canvas from the *Baychimo,* so the captain and a skeleton crew could stay and guard the ship. By mid-October the house was built and the ice was strong enough to support aircraft coming from Nome to evacuate all but the seventeen who were staying with the ship.[18]

The skeleton crew settled in for the winter in relative comfort, with plenty of routine work to occupy them. Some cut up driftwood from the beach, while others hauled blocks of ice from a nearby lake for the water supply. There was a gas engine from a lifeboat to charge the batteries for the wireless. Regularly they sawed and hacked the ice from around the *Baychimo*'s propeller and rudder to prevent damage. Food was no problem, with a year's supply of food on the ship, as well as fresh reindeer meat from the Eskimo at Wainwright. The crew had proper Arctic clothing, plenty of reading material, two gramophones, packs of cards, and many visitors from the local population.

Soon after the middle of November, the sun disappeared below the horizon, and there were only a few hours of twilight during the day. On November 24 the wind increased steadily, and for two days a blizzard raged, keeping the men trapped inside the hut. The horrendous noise from the grinding ice told them that huge changes were taking place outside. When the weather cleared for a short time they discovered that the *Baychimo,* which had been less than a quarter of a mile offshore, had disappeared. There was a mound of ice nearly a hundred feet high where the ship had been anchored, and at first the crew thought it was buried there. But one look proved it was not. No real search could be made until the blizzard ended. The men explored the shore and went as far as possible out onto the ice pack; but finding no trace of the ship, they thought it had broken up and sunk. About a week later a trapper sighted it sailing along on top of the ice pack near Point Barrow, so far offshore that it could not be salvaged. The *Baychimo* was securely frozen into the moving ice field and would continue to sail, sitting high on its ice pan, following the same drift that Amundsen had used to make the first trip through the Northwest Passage in the *Gjoa*. Such was the power of ice, wind, and current in the capricious Arctic Ocean.

The little *Trader* fought its way out through the ice and finally anchored to the ice pan on which the *Baychimo* was riding, looking tiny beside the 230-foot length of the ghost ship. Isobel fearlessly followed the men across the ice pan, scrambled with no concern for dignity up a broken wooden ladder, and was hauled aboard. The hold had contained parkas and curios from the eastern Arctic, but the natives from Wainwright had reached them first. Isobel was later able to buy some of these items in the village. Sacks of mineral ore, caribou skins, sinew thread, writing paper, film, ledgers, even a multivolume edition of the *Times History of the Great War* were there for the taking. Pete and Kari longed to start the engines and salvage the ship, but the *Trader* did not have enough power for the job. They satisfied themselves with taking the gyrocompass. Heavily laden, the *Trader* slowly worked its way back to the distant shore through the twilight and gathering fog bank.

Yielding to the temptation to visit the *Baychimo* nearly ended the eastward voyage of the *Trader,* which would have changed the course of events for Isobel. Unfavorable winds held the little vessel at Wainwright for five more days. They had boarded the derelict ship on the only day when the wind blew the ice pack seaward and opened a lead to Barrow. On that day the *Patterson,* stuck at Wainwright for a month, got around Point Barrow to reach Herschel Island a few days later, using eight hundred pounds of explosive to blast a path through the ice. The *Northland* had seized the opportunity to reach Barrow, canceled the rest of its inspection tour, and headed south because ice conditions were so bad. At Herschel Island the post manager waited impatiently for ships to arrive and noted in his log that they were later than they had ever been.[19] If Isobel was aware of her situation, she gave no hint of anxiety.

Being detained at Wainwright gave Isobel the opportunity to explore ashore and purchase items for Cambridge. "Good to be home!" was her feeling on returning to the warm galley after one of these visits. No longer did its ceiling seem too low; she had learned how to lean back on her bunk till she could stretch out her long legs and sit in comfort in the warm glow from the stove. She knew where to hang the tin plates and cups as she helped Kari wash up each day and could lay her hands on the butter can or the sack of rolled oats at a moment's notice. She played checkers with Pete, while Kari on the bench opposite worked the piano concertina or the gramophone and Ira busied himself with his account book. The promised five-day voyage, lengthening into five weeks, seemed all too short. " 'If you can't get beyond Barrow you can come back with us to Nome and it will cost you nothing,' offered Ira Rank generously." She was almost tempted to say "I'll come," [20] but it was typical of Isobel to reject the easy way and stay with her goal even though the course ahead was uncertain. She had infinite faith in Providence.

Although Isobel was genuinely comfortable in the companionship of the three men on the *Trader,* a letter written to her sister Hilda on notepaper from the *Baychimo* gives an unvarnished picture of her surroundings.

> The three men on board are awfully decent fellows. . . . I'm getting quite handy at knowing how to make stew out of cans. . . . My one regret is a *hot bath*. I am clad in my burberry suit and breeks with fur boots and Bobbie's [a cousin] helmet, but as the stove chimney blows smuts all along the little deck and opportunities for washing anything but my face (and that not often!) are nil, you can imagine me better than I can describe! [21] The "bathroom" is a pail in the engine-room! Golly! I will have lots to tell if I ever get home again. Well, old Ira is snoring in his bunk across the table, the boys are in their bunks in the engine house, and a wild sou'wester is blowing. It is as cold as winter at home. I wear the pony-coat and furboots. I left my other coats at Nome to Mrs. Smith the chambermaid at the hotel, and also sold her my net evening gown for $5. [22]

The final sentence of that letter raises a question. Why was she carrying a net evening gown and presumably lightweight coats on a trip to the Arctic? Possibly the evening gown was used on the outward sea voyage across southern oceans, but it seems unlikely on a freighter with only seven passengers. Possibly it was for the voyage home by ocean liner. Isobel had expected to travel the whole way by ship, and with three trunks she was not traveling light. She had already paid extra airfare for her three hundred pounds of luggage on the flight to Nome. Her choice of clothing was based on the experience of being a resident in an Arctic community in Greenland, but she now found herself in the unfamiliar role of an Arctic traveler with different needs.

The longest part of the journey had been accomplished, and only 110 miles separated the *Trader* from Barrow. With the state of the ice and the unfavorable winds the tantalizingly nearby goal seemed unattainable. After a day of effort they anchored in an ice field dangerously close to the Seahorse Shoals where the *Baychimo* had met its fate. When they were stopped by ice, the men would take the dory and row ashore to search out leads. Isobel went with them and found saxifrage and dandelions blooming amid the cotton sedge and rough blue sea grass along the shore.

Gradually the *Trader* inched forward through the ice, rounding the last point before Point Barrow into Peard Bay. Far out over the ice field they saw the *Baychimo* hurrying northward on its pan of ice, well outside the shoals. Reports of sightings would continue, the last one noted in the Hudson's Bay Archives being in 1969.

Once again the *Trader* was held prisoner by the ice in the inlet of Singnat, where there was a refuge cabin, still twenty-five miles from Barrow. With Pete, Isobel explored inland. Together they cleaned up the cabin, and Pete made a new stovepipe out of gasoline cans, ready to save some traveler's life in the coming winter. The land was flat tundra, but they came upon the mounded ruins of a fairly large Eskimo village and burial site. Another evening, Kari walked miles up the coast toward Barrow, reaching the small schooner of a well-known Arctic trader, Jack Smith, who had news of the anxiety in Barrow over the lack of supplies. The *Holmes,* carrying all the coal for the Barrow hospital, was caught in the ice near the Seahorse Shoals.

During a blizzard on the last day of August, Isobel spent part of the day developing film in the refuge cabin with Kari's help. Three traders from the area, Curran, Henry Chamberlain, and Jack Smith, came along the coast on foot to buy up most of the *Trader*'s stores, concerned that the *Holmes* would not get through. Chamberlain planned to travel east with Gus Masik on his boat the *Hazel,* already in Barrow, and he thought there would be room on board for Isobel. Gus Masik was the former trading partner of Pete Palsson, who had already suggested that traveling with him was a possibility for getting beyond Barrow. The traders along the coast were interdependent and as well known to each other as the captains of the supply boats who came north each summer.

At last, on September 1, the wind blew from the east, opening a lead to the north along the shore. By evening the engines were started as the solid field of ice moved rapidly seaward. They put out a sea anchor at midnight and waited for daylight, but by 4:00 A.M. they were off, terrified lest the wind turn against them. As Isobel regretfully ate her last breakfast on the *Trader,* she saw the wireless aerial, the church spire, and the flag flying from the post office as the boat slipped into the harbor at Barrow.

When Knud Rasmussen reached Barrow in the spring of 1924, it was the largest community he had seen since leaving Godthaab in Greenland three years before. It was then a village of 250 Eskimo and a few non-natives. Point Barrow and Point Hope, on the Bering Strait, were the two places where the Alaskan Eskimo lived by the whale hunt.

The presence of whales attracted outsiders, including Charlie Brower from New York City, whom Isobel met in Nome and now in Barrow. When Brower arrived on the Arctic coast in 1884, the natives were prosperous and independent through their lucrative trading in whalebone. He started the Cape Smythe Whaling and Trading Company at Barrow, married an Eskimo woman, and became the leading person in the area. Although he had been part of the modern influence that caused the Eskimo to abandon the traditional customs governing their annual whale hunt, he fought hard to stop the trade in alcohol and its distillation in the community, where it was having a ruinous effect on the people.[23]

Before she left the *Trader,* Isobel had one important contact to make. Pete Palsson came down the hatchway into the galley with a big, fair man in a drill parka and introduced his friend Gus Masik.

> A pair of very blue eyes sized me up reflectively. "Well," said their owner slowly, "it would sure be tough luck if you had to turn back now when you've pushed on here through so many obstacles. We haven't much of a place for ladies on the *Hazel,* but if you'll take what we've got and don't mind roughing it, you can have my berth in the cabin—there are three down there but it's the driest—and anyway I'll be on deck most of the time. But you'll have to take your chance of being frozen in again round the point. We've got to wait here till the last possible minute for the *Holmes,* and chances are we'll get caught ourselves this year round here."

She accepted his offer on the spot and gratefully took the risk of ice.

"I can promise you that you'll be treated like a lady, and if we make my place before the freeze-up, I'll try to get one of the natives with a motor-boat to get you on to Herschel. If you're lucky you may just do it. If not, I've a big room at my place and you can wait there til the sea freezes and the dogs can travel."

So that was that. The deal was concluded at a price to suit my fast diminishing funds.[24]

Isobel was taken straight to the mission hospital as the guest of Dr. and Mrs. Greist of the Presbyterian mission. There she was given a spotless private bedroom and enjoyed the luxury of a hot bath. But as she saw from her window the riding lights of the *Trader*, she felt a pang of homesickness for her damp bunk and her three friends, Ira (the Watchful), Peter (the Rock), and Kari (the Storm). Those men had become like a father and two brothers to her in the struggle with the ice and the close quarters they had shared. It was a comfortable relationship with real people doing a real job and unlike anything she had experienced before.

Mrs. Greist had a good knowledge of Eskimo artifacts, being a collector for the Smithsonian Institution in Washington and other American museums. She helped Isobel to spend the remaining money from Cambridge and to identify and label the material she had bought along the way, including the ugly pottery bowl from King Island, the only perfect, unbroken specimen of such a bowl Mrs. Greist had ever seen.[25] All was packed ready for shipment on the *Patterson*, expected any day on its return voyage from Herschel Island to San Francisco.

There was great anxiety in the community regarding the fate of the *Holmes*, carrying the coal supply for the hospital, which had only enough left for forty-eight hours. As a diversion for the people, Dr. Greist invited Isobel to give her illustrated lecture on Greenland. On the afternoon of the lecture, word came that the hospital supply ship *North Star*, which had left Nome late in August, had reached the *Holmes*. The wind had changed, the ice had opened, and it was standing by to tow the sailing vessel into port. A rejoicing crowd of nearly four hundred natives filled the church that evening to see Isobel, dressed in Greenland costume, show her slides of Greenland and talk about their distant cousins to the east.

For a few days Barrow was bustling, with most of the northern boats tied up in port loading and unloading supplies as quickly as possible before the onset of winter. Before the *Trader* left to return to Nome, Pete and Kari took Isobel on a tour of the *North Star,* the most modern of the ships that carried supplies for all the hospitals and schools on the Alaskan coast. The *Patterson* stopped in on September 8, picking up the mail for the south. The next morning only the *Holmes* and the *Hazel* remained in port; the *Hazel* would sail that afternoon, taking Isobel on the next leg of her odyssey.

Isobel Hutchison had experienced a journey that few other women had undertaken. Only the handful of non-native women living on the north coast of Alaska would have come in by ship, and almost certainly they would have traveled on the Coast Guard vessel. By 1933 the airplane was changing the means of travel in the North. Native women did not travel far from their own settlements and would have little reason to be exposed to the ice pack. She had made the journey in one of the smallest of the trading boats, had felt the immense power of wind and ice, and had seen its effect on a ship as large as the *Baychimo.* While she was captivated by the icy beauty of the seascape, she was warmed by the genuine friendship of the men on the *Trader,* and that friendship would endure by correspondence for the rest of their lives.

A spirit of comradeship would form between her and Gus Masik in the days ahead as they shared a battle with the elements. It would be hard to find two people of more widely contrasting backgrounds and beliefs, yet their friendship would be another legacy of her continuing journey.

9

Prisoner on a Sandspit

Gus Masik, the man who made it possible for Isobel Hutchison to continue her northern journey, was not only honorable and tough but also immensely capable. He was born on a farm in Estonia in 1888, making him six months older than Isobel. Wild as a youth, restless and freedom loving, he ran away to sea in 1905 at the time the Russian Revolution was creating havoc in the Baltic states. Surviving the brutish life of an ordinary seaman in half the navies of northern Europe, he eventually reached America and then Alaska. Physically strong and able to turn his hand to almost anything, from mining to boat building, Gus was never without work. In the course of his life in the North he worked for Vilhjalmur Stefansson, leader of the Canadian Arctic Expedition, and when that was finished he traded and trapped up and down the Siberian coast, risking death or imprisonment during the conflict between the Red and White Russians in 1917.

Gus had the potential to become a leader in a northern community, but two conditions stopped him. If his work was too easy or if he had too much money in his pocket, his restless nature drove him to move on. In 1933, when Isobel met him, the rolling stone was living in a one-room cabin at Martin Point on the shore of the Arctic Ocean, operating as an independent trader, trapper, and prospector.[1]

There was a sense of urgency when Gus Masik arrived at the mission hospital to collect Isobel for the journey east. Loading had proceeded day and night from the moment the sailing ship *Holmes* arrived in port at Barrow, and Masik's boat *Hazel* was now anchored two miles up the coast away from the ice, ready to leave. Barrow is on the west side of Point

Barrow, and this most northerly projection of the United States was a point of anxiety to all sailors traveling eastward.

Though about the same size as the *Trader,* the *Hazel* had a larger cabin where Gus could stand upright. That advantage was balanced by the cabin's being also the engine room, noisy and smelling of oil and gasoline. The hold and decks were loaded to capacity with gasoline cans, oil drums, lumber, and trade goods. The frozen carcass of a caribou hung from the rigging and supplied breakfast, lunch, and dinner for each day of the voyage.

In addition to Gus Masik, the crew consisted of the engineer, Henry Chamberlain, whom Isobel had met when the *Trader* was caught in the ice west of Barrow, and two Eskimo, Tigutak and Niniuk. Gus and Henry occupied the bunks on the side of the engine opposite Isobel, and the two Eskimo slept amid a pile of fur in the hold. Towed behind the *Hazel* was a shallow-draft motorboat with a canvas cover, loaded with trade goods for Flaxman Island and carrying two Eskimo girls and their father, Samuel.

After rounding Point Barrow, the *Hazel* anchored for the night while the native boat continued eastward through the long lagoons that fringed the shore. The next day, September 10, dawned windless and fair, but already the open water was ominously crusted with young ice. The *Hazel* continued, crossing Smith Bay and picking up the native boat farther along. Other traders' boats were following in their wake, all racing the ice to get home.

> "How many people would give their ears to be where you are to-day!" said Gus to me next morning, when, breakfast over and my self-imposed task of washing up done, I had climbed on deck to sit under him on one of the gasoline boxes by the steering-wheel.
>
> "You will have a lot to tell when you get home again," he cried; "not many have seen what you are seeing!"
>
> I suddenly realized with a rush of wonder at circumstance, how true his words were.[2]

By the end of the second day they were halfway to their destination and anchored in fog at Thetis Island. It was here that Stefansson and some members of the Canadian Arctic Expedition had made their way ashore from the icebound *Karluk,* the ill-fated ship that drifted a thousand miles westward and sank near Wrangel Island. One of the men, the ethnologist Diamond Jenness, had spent several months along the stretch of coastline they had just passed, living with Eskimo families, learning their ways, and transcribing their stories as part of his work for the expedition.[3]

The *Hazel* was able to continue through the shallow lagoons between the chain of low islands and the coast. At Beechy Point they called in at Jack Smith's trading post, run by Aarnout Castel, Gus Masik's old friend and partner from both the Canadian Arctic Expedition and their Siberian venture. Castel was married to a native woman who had been "outside" and spoke good English. Isobel enjoyed the break from the boat, spent in the pleasant kitchen of their home while the men unloaded trade goods. Delays in this season were dangerous, and within two hours the *Hazel* was ready to leave. The Castels and their three children piloted the *Hazel* out across the bar of the lagoon to wave good-bye from their motor launch. At that moment the isolation of the area was borne in on Isobel as she realized no other boat would call there for nearly a year.

The following day they stopped at Flaxman Island, where Henry Chamberlain, the engineer on the *Hazel,* had his trading post in a small native settlement. Unloading went on well into the evening, giving Isobel a chance to stretch her legs ashore. The snow was knee-deep in places as she walked along the edge of the lagoon. The women she met on the beach all shook hands with her, but communication was limited to smiles when the Greenlandic words Isobel tried were not understood. She was told later that she was the first white woman many of them had seen.

Whether unconsciously or by design, Isobel Hutchison used her encounter with the Eskimo women at Flaxman Island to help her readers appreciate the uniqueness of her experience and to emphasize that she was almost certainly the first non-native woman to have traveled along that coast. The isolation of the Castel family and her quoting Gus's remark about how few people had seen what she was seeing further underscored the point. A travel book should contain material that is not only interesting but exceptional, and though she never exaggerated hardships, she occasionally drew attention to the remoteness of her travels.

Flaxman Island was home to the native boat they had been towing and to Tigutak, one of the helpers on the *Hazel.* The boat became the scene for a sociable evening when Mrs. Tigutak came aboard to cook supper for her family and the crew while Isobel was ashore. At this point on the coast, Isobel was treading ground that had connections with Arctic explorers she had already met. It was at Flaxman Island that Ejnar Mikkelsen wintered in 1906 with Ernest Leffingwell before they explored the sea ice the next spring. They were befriended at Flaxman Island by an Eskimo family who had left the relatively crowded conditions at Point Barrow to seek freedom in the empty spaces of the coast.[4] Leffingwell stayed on at Flaxman Island for several years, mapping the coast and exploring the Endicott Mountains, and he was still in the area when Diamond Jenness and members of the Canadian Arctic Expedition arrived in the spring of 1914 on their way east.

Knud Rasmussen, in search of original ethnographic material in 1924, had not deemed it worthwhile to spend any time between Demarcation Point, the border with Canada, and Point Barrow, since the natives there had long been in contact with outsiders. "The Eskimos are scattered about in little encampments all along this coast; we find, too, a few white men, Scandinavians for the most part, some with small schooners, others with nothing but their bare hands and their traps. The distance between dwellings depends on the chances of a good haul."[5]

Lacking the dramatic cliffs of the western coastline fronting on the Bering Strait, the area east of Barrow is particularly bleak, with shallow lagoons behind sandspit islands. "Flat, low and dreary" were the words Hudson Stuck used to describe it on his circuit of the Arctic coast, saying it was a place where few people lived, since the Eskimo migrated inland to follow the caribou.[6] In spite of that, Isobel was captivated by the peace and beauty and was able to accept her extraordinary situation with equanimity. She entrusted her life to God and Gus Masik and had complete faith in both.

With a west wind, the *Hazel* hoisted sail and, crossing Camden Bay, passed Barter Island without stopping at the trading post of Tom Gordon, a Scottish whaler. From a distance Isobel noticed that his house resembled a Hebridean farmhouse. At last they rounded a long, narrow sandspit almost surrounded by the sea and entered a lagoon about a mile wide that divided the spit from the mainland. " 'Well, here we are! Home at last!' says Gus, with relief in his voice, as he turned from the steering-wheel, flapping his benumbed arms."[7]

It was September 15, and with favorable weather they had made the journey from Barrow to Martin Point in just six days. Only 120 miles still lay between Martin Point and Isobel's desired destination of Herschel

Island. Simon, the Eskimo who might have been willing to convey her there in his motorboat, had gone west to meet the *Hazel* and had passed unseen in the mist. When he returned four days later, he had difficulty getting through the ice to reach the sandspit, and everyone knew the season for boating was over. At the Hudson's Bay Company post on Herschel Island, the last trading boats had left and the thermometer registered $-22°$F.[8] Winter was at hand. "I was a prisoner upon a snow-covered strip of shingle about a mile long and scarcely in any part more than a hundred yards wide, washed on all sides by the sea. Escape—had I desired it—was impossible after the first few days, during which the lagoon between the Sandspit and the mainland was still open."[9]

Arriving at Gus Masik's cabin, locked in by the ice, Isobel was in a compromising position for a single woman traveling alone. She had only Gus's word, given when they met, that she would be treated like a lady, but in no way had he undermined that promise during the journey. All her life Isobel had sought deep and abiding friendship, but apart from her friend Medina Lewis, from college days at Studley, it was rare for her to find such empathy among her ever-widening circle of acquaintances. Already, from her observations during the week of travel on the *Hazel,* she felt a growing affinity for Masik. Soon after her arrival on Sandspit Island, Isobel's diary records a memo for the month of September: "A new interesting month on board *Hazel* and at Gus Masik's. A new life for me learning to be a trapper and trader in the Arctic and getting life story of a new kind from GM. Have had a very happy month and have found a real new friend in Gus."[10] Isobel could admire his competence and strength, but it was the flame of independence burning in his spirit that matched her own. Gus had never met a woman like Isobel, and their regard was mutual.

Sandspit Island contained two dwellings. One was a small, untidy, aromatic Eskimo house made of driftwood, and the other was the quaint "round-house" of wood, turf, and canvas that was Gus Masik's home and trading post. The Eskimo house was occupied by Nokipigak (known as Bruce), his wife Jenny, and their four children. Gus's house was clean, cozy, and well built, lit by two skylights. It was approached through a canvas-covered entry and storehouse, with wooden compartments for the four sled dogs. The subject of suitable accommodations for Isobel until freeze-up made further travel possible was settled without hesitation. "The question of where to put his lady-boarder in a single-roomed cabin might have worried some men farther south, but it did not trouble Gus. The tentative suggestion that I might set up my camp-bed in the canvas-roofed entry with the dogs brought a stentorian roar of—'What! And freeze to death! *Nothing doing!* Isn't this room plenty big enough?' "[11]

By now Isobel had learned in her travels to accept the available accommodations with good grace and make the best of it. She set up her camp bed in the corner opposite Gus's wooden bunk. Two nails driven into the roof provided pegs so her traveling tent could be hung at night as a privacy curtain.

While they were traveling on the *Hazel,* the noise of the engine had limited conversation, and Gus was fully occupied piloting the boat through ice and shallows and overseeing the unloading at stops along the way. Having arrived on Sandspit Island, Gus and Isobel had time to take stock of one another:

"You took an awful chance travelling alone in these parts! . . ."

"I never take any chances," I said, quite sincerely. "God always blazes my trail. I should not be here now if He hadn't."

"You really believe that?"

"Absolutely."

"God! Why, that's all bunk! I don't believe in God. I believe in the five-year plan. Didn't I tell you I was a Bolshevik?"

"Well, I'm a Christian, and I believe in Eternity."

Gus collapsed with a roar of laughter.

"Well," he confessed handsomely, "you're the first *lady* I've ever met, I guess."[12]

In italicizing the word "lady" in her book, Isobel drew attention to the strategy she adopted for keeping a safe distance between them in such close quarters. An example of their domestic arrangements appears in her diary: "Oct. 1. Had a bath—Gus fixed it up for me and then took a walk! So we do the best we can in our double menage. . . . Oct 3. Went for a walk on snowshoes while Gus had a bath."[13]

For the first week Gus was busy unloading the *Hazel,* preparing for Henry Chamberlain to take the boat back to Flaxman Island for the winter. Gus spent many days arranging his stores, hunting seals for food for his four dogs, and adding to the woodpile. The immense amount of driftwood floating westward from the Mackenzie River was the reason he had chosen to build his cabin on the sandspit. While the water in the lagoon was still open, Gus ferried Bruce and his family with their tent to the mainland to go hunting. Isobel enjoyed the chance to walk a short distance across the tundra and saw carex grass and a single yellow poppy coated in ice showing through the covering of snow.

When the weather was favorable, Isobel walked each day to the end of the sandspit, which reminded her of Tiree, one of her favorite places in the Inner Hebrides. She assumed that the sled dogs needed exercise, like her dogs at home, and took each one in turn to the end of the spit. To Gus's great amusement, they almost pulled her arms from their sockets. Gus warned her to watch for polar bears and also to keep an eye on the weather: a blizzard could come up very quickly.

One day John Olsen, Gus's former partner, now living east of Martin Point, came to collect his annual mail, a single parcel sent from Nome and brought from Barrow on the *Hazel*. He was the nearest neighbor, and they arranged to pay him a visit when sledging became possible. The opportunity came early in October when the ice on the lagoon was solid, and they passed his cabin on the way to visit Mrs. Ary, one of the few women with traditional skills living along that sparsely inhabited coast. She had been engaged to make Isobel a fur outfit for the sled journey east.

> Mrs. Ary was a gentle-faced old Eskimo woman of the old style, her chin tattooed with three blue stripes after the fashion of former days. She and her family lived in a typical native house of driftwood. Passing through an outer entrance piled with lumber, old cooking pots, frozen fish, and what not, the living-room was entered through a wooden door in which was a minute window some two inches by three—of glass. The room was furnished with a few bed-platforms piled with skins (on these I had been warned by Gus *not* to sit) and with a large iron stove. It was lit by a window of gut-skin at the far end.[14]

Mrs. Ary spoke no English, although she was the widow of Ned Arey (or Ary), an American whaler who arrived in Barrow in 1887 and became one of Charlie Brower's partners. Diamond Jenness had received hospitality from her parents, the Kunaluaks, twenty years earlier.[15]

On her initial visit to Mrs. Ary, Isobel was measured for a reindeer skin parka and trousers. She returned a week later for a fitting, walking the seven miles east by herself. On her way back Gus met her with the dogsled at John Olsen's cabin. In payment for the work, Mrs. Ary had requested a set of white woolen underclothes, some three-cornered needles, calico, drill, and other material. All of these things, including the reindeer skins, Isobel was able to purchase at Gus's trading post.

The solo visit to Mrs. Ary is an indication of how comfortable with her surroundings Isobel had become during her first month at Martin Point. She was accustomed to taking long walks alone, and if she followed the shoreline there was little danger of losing her way. Knowing that in the Ary household neither her English nor her Greenlandic words would be understood did not deter her. Gus had already warned her to watch for polar bears on the sandspit, and he no doubt repeated the warning as she set off. Concern and friendship are shown by his going partway to meet her with the dogsled, and undoubtedly she kindled his admiration with her independence.

While Isobel was enjoying the unruffled pace of life on Sandspit Island, taking long walks, finding flowers in the snow, and visiting her seamstress, the Canadian authorities at Aklavik and Herschel Island were sounding the alarm. She was known to have left Barrow on September 8 and nothing further had been heard of her. On October 3 the police schooner from Aklavik arrived at Herschel Island, making an unprecedented trip so late in the season.[16] Because she was a woman, and thought to be traveling alone, Ottawa urgently requested that Inspector Charles Rivett-Carnac investigate Isobel Hutchison's whereabouts and pick her up at Herschel Island if she had arrived there. He reported:

> I was in some doubt as to the necessity of this patrol being made and in the case of anyone of the masculine gender would have considered that any difficulty they might get into to be their own responsibility if they cared to hazard the Arctic Coast at that season of the year, but taking into consideration Miss Hutchison's sex and the fact that I understand that this lady has been assisted to a great extent by the Danish Government in Greenland on her excursions in that country, I considered that the proper course which would be approved would be to help her as much as possible in a similar manner, especially in view of the fact that the journey she was undertaking, while safe enough if travelled by experienced Northern men, is no easy matter for an unattended lady to complete, and might be regarded as containing a certain amount of hazard unless made by a person acquainted with Arctic Coast conditions.[17]

In that long sentence Inspector Rivett-Carnac summed up the precarious nature of Isobel's journey and showed that she was viewed as a person of some importance. Gender was the issue that tipped the balance and sent him on an inconvenient journey. A man could be allowed to suffer the consequences for getting into difficulties through his own carelessness in the Arctic, but chivalry demanded that a "lady" be rescued. "An unattended lady" was a lady in distress. Although all women belonged to the gentle sex and should not be roaming by themselves in the North, by referring to her as a "lady" Rivett-Carnac gave her a class status that implied important connections. As the representative of the paternal state that was responsible for everything that happened in the North, he could not ignore her "plight."

The inspector was extremely reluctant to make the trip because the birth of his first child was imminent and his wife would be alone. But because it was so late in the season and navigation would be difficult, he felt it essential that he take charge of the patrol himself.[18] Should they hit bad weather and the schooner become frozen in at Herschel Island, the men would have to return to Aklavik on foot, a particularly difficult journey in the season before lakes and rivers were solid.[19]

The concern of the Canadian government for locating Isobel Hutchison contrasts with the United States government's lack of interest in her movements within Alaska from the moment she crossed the border on the steamboat *Yukon.* Alaska had achieved territorial status in 1912 with limited power, but government service was thinly spread over the area, one-fifth the size of the United States and physically disconnected by the intervening country of Canada. On her journey around the coast, Isobel had seen little evidence of the American government. In Point Hope and Wainwright there were schools for native children, underfunded by the Bureau of Education and sponsored by various church denominations. In Barrow the relatively new hospital provided by the government was staffed by the Presbyterian Church. When circumstances required direct action in Barrow, Charlie Brower, as the white resident of longest standing, was often designated to act on behalf of the government.

In the matter of law enforcement there was stark contrast to the situation just across the border in Canada, where the Mounties were the dominant arm of government. Just the previous winter a Mountie, going to a trapper's cabin to investigate a complaint, had been shot and killed by the man who became known as the Mad Trapper of Rat River. The trapper fled and was tracked for two months by a posse of police and an airplane before he was cornered deep in the Richardson Mountains and shot. It is now believed that the trapper was an American fugitive from one of the notorious gangs of outlaws that operated in the western states.[20] In Alaska law enforcement was carried out by the annual cruise of the revenue cutter, which in 1933 had been prevented from going east of Barrow by the ice. As Gus Masik frequently pointed out to Isobel, even serious crimes went undetected and unpunished in that corner of Alaska. In these circumstances it was not surprising that the American government ignored the rambles of a whimsical botanist.

Not finding Isobel at Herschel Island and with no information on her whereabouts, there was nothing the Canadian police could do. They quickly turned around and made the three-day trip back through a ferocious storm on the coast and up the Mackenzie River with the ice closing in behind them. Two weeks later the post manager at Herschel Island recorded a blizzard from the northwest, the heaviest he had ever seen.[21]

On Sandspit Island the same storm raged for two days, making it impossible to go outside. "This storm was followed—as usually happens—by a still sunny day of glorious calm. The north side of our sandspit island presented an astonishing spectacle, being piled with great square blocks of ice—the 'pressure-ridge' of ocean ice—which was crushed up to a height of twenty feet or more all along the outside of the spit."[22]

Inside the one-room cabin Gus maintained a sense of order that allowed each of them to operate independently. Both spent as much time as possible outside, and they both had sufficient inner resources to occupy themselves. When confined to the cabin, Isobel mounted and labeled the seven hundred specimens of plants she had collected along the way and developed photographs, while Gus repaired the harness for the dog team or built new shelves for the trade goods. Gus wisely took charge of the cooking and breadmaking, and Isobel did her share by washing the dishes. The two of them enjoyed many friendly arguments on religion, Bolshevism, capitalism, and politics over the ensuing weeks. In discussing belief, atheism versus Christianity, the argument often led to Gus's contending that Christianity had destroyed the life of the native people, a subject Isobel did not have the experience to refute. The only source of irritation for Isobel was Gus's habit of filling the cabin with tobacco smoke in the evenings, but she was conscious of being a paying guest and rarely complained.

On days when they were confined to the cabin, and during the lengthening evenings, Isobel encouraged Gus to tell her the story of his life as a distraction for them both.

As Gus continued his stories most evenings, Isobel transcribed them in his vernacular, screening out the frequent expletives. The result was published as *Arctic Nights' Entertainments: Being the Narrative of an Alaskan-Estonian Digger, August Masik, as Told to Isobel Wylie Hutchison.* The review in the *Times Literary Supplement* praised it as "an excellent piece of reporting, unweakened by literary artifice. Any one of Masik's adventures would furnish a rich platter of pseudo-epic narrative. Miss Hutchison leaves the dish as he served it. There is nothing here but the hard crust of adventure and the rough salt of the man."[23]

Through Gus's stories Isobel knew that she was living with a real northern adventurer, and also through him she understood the intricate web of connections between most of the non-natives who traveled that coastline. During his stint with the Canadian Arctic Expedition, Masik was part of a small group that made the longest unsupported trip on an ice pan in the Arctic Ocean up to that time. Storker Storkersen, who had been with Ejnar Mikkelsen on a similar trip in 1907, was the leader. Stefansson, whom Isobel would know only through letters, had written of Masik in a letter that Gus proudly carried on his person: "There is no man who has been with me in my Arctic work in a non-scientific capacity whom I would rather have with me again."[24]

Isobel was equally fascinated by the stories of Gus's exploits in Siberia. She had heard something of these on the *Trader* from Ira Rank and Pete Palsson, who were part of the "mosquito fleet" of little boats that dared to cross the Bering Sea from Nome when trading was permitted. Siberia was high on Isobel's list for future travel, as she had already demonstrated by applying to the USSR for permission to go there.

Toward the end of October Isobel received an invitation to visit the old Scottish trader Tom Gordon, whose house on Barter Island had caught her eye on the passage from Barrow. It was on Barter Island that Stefansson and his group, including Gus Masik, wintered before their ice trip in 1918. It was on this island that Diamond Jenness made extensive archaeological excavations of the remains of early Eskimo houses on the sandspits on both ends of the island.[25] With the wind behind them, Isobel, John Olsen, and one of Gordon's sons made the twenty-five-mile trip by dog team in two hours.

Like Gus Masik, Tom Gordon had run away from home to follow the sea and never returned. His forty years in the North began when the ice pack forced ashore the whaling ship on which he was captain. He teamed up with Charlie Brower at Barrow working for the Liebes Fur Company of San Francisco and eventually married an attractive Eskimo woman. In 1914, when the Canadian Arctic Expedition wintered at Martin Point, Gordon ran a trading post for the Liebes Company near Demarcation Point, just west of the Canadian border.

A vivid picture of Tom Gordon and his home is given by David Irwin, the young man who traveled to the Arctic coast on the *Trader* the year before Isobel Hutchison made her trip. Irwin spent some time working for the reindeer drive from Nome to Canada that had been in progress for several years and was currently passing a few miles south of the coast.[26] He had come to Tom Gordon for help when things went badly for him and was treated with generosity and kindness. He described Gordon as a giant of a man, dressed in a blue flannel shirt, deerskin pants, and mukluks, with shoulder-length white hair, smooth-shaven red face, and keen, flashing blue eyes under fierce tufted brows. Like a feudal king, firm and just, understanding and kind, he ruled his dynasty of fourteen sons and daughters, some of them now married with families of their own.

The low, sprawling house was built of many types of material—clapboard for part of the walls and roof, with blocks of sod eighteen inches thick laid against a framework of spars and driftwood over the rest. It contained two main rooms, and Gordon and his wife lived alone in one of them, following white custom. Their room was low ceilinged and wide, the inside studs papered with cardboard from cartons. Pegs held clothing, harness, rifles, and gear of every description. On a wide table in the center was a shaded kerosene lamp; chairs cut out of barrels and a roughly made bunk completed the furnishings, and there were two polar bear rugs on the floor. The rest of the large family lived in a lean-to built of sod blocks at the other end of a connecting passageway, separated by a calico curtain. They lived in native fashion, eating on the floor; each family had its own corner for clothing and sleeping bags, but they shared their food. Trading was also carried on in the family's room.[27]

When she sailed past in the *Hazel,* Isobel had seen Tom Gordon's house as a Hebridean farmhouse, but the reality did not match her romantic image. In her diary entry about the visit Isobel referred to the Gordon family as "the queer Scotch-Eskimo household."[28] This is perhaps a subtle hint of discomfort with miscegenation, although Isobel had accepted without question other mixed marriages she encountered both in Greenland and on the Arctic coast. Within his marriage Tom Gordon had "gone native," yet he appeared to live apart from his native family as a Scot, emphasizing the split nature of the union. "The time was passed in the room occupied by Mr. Gordon and his wife, next-door to the capacious kitchen which housed the rest of the family—Mr. Gordon pacing to and fro as if he were still on the deck of his ship, waving a constant cigarette and describing to me some of the adventures of his youth."[29] Isobel had expected to stay for one night only, but a blizzard held her there for a second one before she was able to return to Martin Point.

Back on Sandspit Island, winter darkness was closing in. Early in the morning on the last day of October, Gus roused Isobel, telling her to put on her parka and come outside to look at the moon. "It was truly an astonishing sight, sinking over Barter Island on the western horizon, blood-red and almost square, as large as a rising sun. To eastward the sky on the horizon line was a faint rose with saffron shading to blue above, and a star or two still clear in the emerald zenith."[30]

The full moon of November 3 set the date for the 120-mile journey east to Herschel Island. News traveled swiftly on foot along the coast. On Herschel Island it was known that Isobel Hutchison was six days west and preparing to travel.[31] At that time of year the few persons traveling along the coast were welcomed at every outpost as a source of news. The manager may have been able to pass along the information to the police at Aklavik, but it was so late in the season that there was nothing they could do. Bruce, the Eskimo who lived on Sandspit Island, had been promised $80 for the dogsled trip, but he had not returned from his hunting camp on the mainland. Rather than delaying, Gus cheerfully chose to conduct Isobel to Herschel Island himself.

In Isobel's description of the early morning departure there is more

2. Carlowrie. Photo by Alex Hoyle.

1. (*Overleaf*). The young family at Carlowrie in 1893. Hilda, Walter, Isobel, and Nita. Photographer unknown.

3. Knud Rasmussen on his boat, *Nanortalik,* 1927. Photo by Isobel Wylie Hutchison. Courtesy of Trustees of the National Library of Scotland.

4. The mountainous island of Umanak. Photo by Isobel Wylie Hutchison. Courtesy of the Royal Scottish Geographical Society.

5. Danish houses and Greenland igdlos, Umanak village. Photo by Isobel Wylie Hutchison. Courtesy of Trustees of the National Library of Scotland.

6. Isobel Hutchison in new Greenland costume. Photographer unknown. Courtesy of Trustees of the National Library of Scotland.

7. Dr. Morten Porsild, scientist, and his wife at their
house in Godhavn. Photo by R. McKeand, D.L. Courtesy
of the National Archives of Canada. PA-138962.

8. A Point Hope Eskimo at the door of his
home. Photo by Isobel Wylie Hutchison.
Courtesy of Gresham Publishing Company,
New Lanark, Scotland.

9. The *Trader*, Ira Rank in front, Kari Palsson with trading Eskimo behind. Photo by Isobel Wylie Hutchison. Courtesy of Gresham Publishing Company, New Lanark, Scotland.

10. Isobel Hutchison perched on the bow of the icebound *Trader*. Photographer unknown. Courtesy of Trustees of the National Library of Scotland.

11. The little *Trader* anchored to an ice floe beside the *Baychimo*. Photo by Isobel Wylie Hutchison. Courtesy of Gresham Publishing Company, New Lanark, Scotland.

12. "Natives of Wainwright with loot from *Baychimo*," 1933. Photo by Isobel Wylie Hutchison. Courtesy of Hudson's Bay Company Archives, Provincial Archives of Manitoba. HBCA 1934/2/6.

13. Gus Masik's cabin on Sandspit Island. Photo by Isobel Wylie Hutchison. Courtesy of Gresham Publishing Company, New Lanark, Scotland.

14. Gus Masik with his lead dog, Whitie. Photo by Isobel Wylie Hutchison. Courtesy of Gresham Publishing Company, New Lanark, Scotland.

15. Isobel Hutchison at Demarcation Point, one foot in each country. Photo by Gus Masik. Courtesy of Gresham Publishing Company, New Lanark, Scotland.

16. Erling Porsild on skis at the Reindeer Station house. Photographer unknown. Courtesy of the National Archives of Canada. PA-130437.

17. Pilot McMullen refueling at Fort Resolution during the flight "out." Photo by Isobel Wylie Hutchison. Courtesy of Gresham Publishing Company, New Lanark, Scotland.

18. Chief Mike Hodikoff and his family on Attu: Isobel Hutchison, back row. Photographer unknown. Courtesy of Gresham Publishing Company, New Lanark, Scotland.

19. "Admiral Hutchison of the Bering Sea" with the officers of the *Chelan*. Photographer unknown. Courtesy of Gresham Publishing Company, New Lanark, Scotland.

20. Gus Masik with Isobel Hutchison at Carlowrie, 1937. Photographer unknown. Courtesy of Trustees of the National Library of Scotland.

than a hint of regret at leaving. "I passed for the last time through the little wooden door with its fringe of reindeer fur. I had left my comfortable home of nearly seven weeks for good, and turned my face to the Arctic snows."[32] Once again she was stepping off into the unknown. At least for the journey to Herschel Island her companion was a proven friend.

The first day led past the homes of John Olsen and Mrs. Ary. At Mrs. Ary's house her grandsons were ready to depart for their home farther east. They lent Gus one of their dogs to help pull the sled, heavily loaded with Isobel's mountain of luggage. The trail descended to the sea ice, rough with heaped cakes thrown about by tide and wind. Both groups camped together in an abandoned cabin about halfway to Demarcation Point after making thirty-five miles the first day.

Fog enveloped them soon after their departure the second morning, hiding not only the trail but also the view of the Endicott Mountains, closer now as they traveled east. The Arys, with lighter loads, went ahead to their home, leaving a trail Gus could follow. On this lonely stretch of coast Isobel and Gus stopped at a little group of cabins, home to both Paul Ary and a son of Tom Gordon, to get directions for the trail ahead. They learned about an abandoned Hudson's Bay post at Demarcation Point, but an empty Eskimo house before the border made a more comfortable shelter. The Eskimo were away at their hunting camp, and in the custom of the North, travelers were welcome to use what they needed.

In sunshine on the third morning, Gus went ahead to pick out the trail through a rough entrance to the sea ice, leaving Isobel to manage the dog team down the steep bank from the tundra. Forgetting to use the brake, she let the dogs careen down the bank out of control, tipping the six-hundred-pound load onto the ice. Swearing at the dogs, Gus uttered his first reproach to Isobel: "That's all the help *you* are on the trail." With their combined effort they righted the sled without unloading it, but Gus's toe was crushed between the runners and the ice. As he took off his mukluk to check for broken bones, Isobel realized how completely helpless she would be in that desolate region without her guide. The injury was not serious, but it caused Gus to limp for some hours. At the obelisk marking the border Isobel used the last of her film, and Gus celebrated with a shot of "moonshine." Isobel believed she was the first non-native woman to enter Canada at Demarcation Point.

A smooth sheet of sea ice helped their progress, and they passed the abandoned Hudson's Bay post, now occupied by an Inuit family. Gus was anxious to reach the next house at Kamarkak, around the point of the bay, but with the setting sun the wind began to gust, and it increased in force as darkness grew. Gus inspected two deserted houses on the shore, but pronounced them unfit for habitation. They set up Isobel's tent, and she cooked rice and raisins over the Primus stove while Gus built a proper igloo for the night. Isobel ignored Gus's comments that missionaries had so destroyed the natives' way of life that none of them were now capable of building snow houses. The day had been too strenuous for her to enjoy an argument.

"Well, are you going to stay in the tent and blow away, or do you want to share my house?" I went out into the blowing night. Banners of green light were threading in and out across the clear starlit sky, their weird tentacles clutching now here, now there; low in the west hung a splinter of moon; amid the snow and ice-bergs that covered the sea rose the fairylike dome of a snow-house, a light from within glimmering through its walls till they shone like the hood of a daffodil in sunshine. Crawling on hands and knees through the low doorway, I found myself in a tiny crystal chamber, its wall, floor and roof shining like diamonds in the light of a candle frozen to the floor. My choice was easily made. "I'll share yours!"[33]

After some rough going over the sea ice on the fourth morning, they reached the house at Kamarkak and were shown the route over the lagoons where the trail turned inshore. Thick fog on a windless day created a penetrating cold. They were glad to reach a hunting cabin where two boys, ahead of them on the trail, already had the fire lit. "Gus acting as my cook for the last time, prepared a savory stew for everybody with the remains of our reindeer meat."[34] As she moved too close to the fire to stir the stew, the tails of Isobel's burberry jacket caught fire.

In referring to Gus as "her" cook and "as cook to dogs and lady," Isobel jokingly turned a blind eye to the incomparable service he had just provided in conducting her along one of the loneliest stretches of the Arctic coast. By emphasizing the class difference between them that neither ever forgot, she minimized the comradeship they had experienced for the past two months. In her book *North to the Rime-Ringed Sun* she lightly passed over her friendship for this remarkable man. Besides her early comment that she had found a real new friend in Gus, in her diary at the end of October she noted "October in G's little round house at the top of the world, very happy month of Arctic life, learning to be an Arctic housekeeper."[35] Her thoughts were on the reading public as she used the strategy of signaling a lady-master/servant relationship. She was walking a fine line in writing about their unconventional living and traveling arrangements, trying to ensure that her words would not be misconstrued.[36]

The boys, also going to Herschel Island, were traveling with an almost empty sled. Gus paid them to take two of Isobel's heavy boxes, and the two teams raced through the fog toward Herschel Island, arriving there before midday. "A tall dark young man stepped forward as Gus drew up his team. 'Miss Hutchison, I presume?' he said. 'The police-boat from Aklavik was here in the beginning of October looking for you, the latest run they ever made up here! We didn't know where you were, except that you had left Barrow.' "[37]

With the shipping season finished, news traveled along the coast by "moccasin" telegraph only. Word of Isobel's having left Barrow would have come as an exchange of coastal gossip with Captain Pedersen of the last ship of the year to reach Herschel Island from Barrow. In 1933 there was no wireless telegraph along the north coast, and the only radios were short-wave receivers bringing in news from far to the south. Isobel was amused by the concern about her whereabouts. She had embarked on her journey as a private person with no expectation that any government agency would be looking out for her welfare. In all her travels so far, in Iceland, in Greenland, and in Alaska, she had operated entirely as a free agent.

Two days later Isobel shook hands with Gus, bade farewell to his beautiful dogs, and watched wistfully as they skirted the headland and disappeared into the wilderness of ice and snow that was their home. An enduring friendship had formed between this unlikely pair, built on mutual respect and trust. Gus, capable of anything he tackled, honest and honorable, gruff and thoughtful, respected the unwavering courage of gentle-born Isobel, who could match wits with him in any argument. Each had expanded the other's horizons, Gus with his tales of knocking about the world and his involvement in Arctic exploration, Isobel with her love of nature, her knowledge of plants, and her store of folklore, poetry, and Scottish legend. They would meet twice more, in Nome and at Carlowrie, and their letters would continue until Gus's death in 1976.

The journey to Aklavik was still ahead, a trek of 150 miles from Herschel Island. On that part of the trip Isobel would have to arrange for new dog teams and drivers at each stop and would miss Gus's steady companionship. Surprisingly, when she reached Aklavik she would also miss the lack of regulations that had allowed her to be an independent traveler in Alaska, with her arrangements falling into place as circumstances dictated. Having arrived in the Canadian Arctic on her own initiative, under the aegis of no institution, she was suddenly the responsibility of the Mounties and the government, and they were not accustomed to dealing with a free spirit.

10

By Dogsled to Aklavik

Herschel Island, eight miles long and no more than four miles wide, had many ships in port during the summer as the meeting place of the eastern and western Arctic. When Isobel arrived on November 7, 1933, the trading ships had been gone for two months, and the population of the island had shrunk to its normal complement. For ten months of the year, Special Constable Ethier of the RCMP and his wife, James Sinclair from Orkney, the post manager for the Hudson's Bay Company, and two native families who worked for the company were the only people living on the island.

With the best natural harbor in the area, Herschel Island had a shorter but more intense involvement with the whaling trade than did the Alaskan coast. The first American whaler reached Herschel Island in 1889, and for the next few years ships that penetrated so far east from the Bering Strait tended to winter there. With the whaling crews idle for months at a time on a small island, the native people experienced even more aggravated effects from the contact than those on the Alaskan coast.

Nuligak, an Inuk who spent part of his childhood on Herschel Island, remembered its many buildings: houses, stores, a workshop, and a watchtower with a foghorn. Whaling crews were a mixed lot, Alaskan and Siberian Eskimo as well as white sailors shanghaied from southern seaports. It was not unusual for the white sailors to try to escape south through the mountains during the winter. "The sailors of the whaling boats were rough and tough and not all of the best quality," Nuligak reported. "An old Eskimo woman once spoke to me about her father, a sailor on a whaling vessel: 'My father was fervent in his prayer—but no one could beat him when it came to wine and women.' "[1]

Being so far from the regular travel routes, the foreign presence was unnoticed by Canadian authorities until the missionary Isaac Stringer, making his first visit to the Arctic coast in 1894, found the Eskimo debauched with alcohol and syphilis. That winter there were twenty-two whaling ships wintering at Herschel.[2] The government moved quickly to establish a Canadian presence on the island with a Mounted Police post. The decline in the whale population from overfishing and the collapse in the price of whalebone in 1907 gradually returned Herschel Island to the quiet place it had been before the whalers discovered it.

The island was wrapped in several feet of snow when Isobel arrived, and she was warned not to wander far from the now deserted village because of the danger of sudden blizzards. Constable Ethier and his wife provided hospitality in their home, and Isobel felt strangely unsettled by the return to civilized comfort. With the help of Mr. Sinclair, she developed and printed photographs.

Although she had written from Nome to the director at Kew that she might have to winter on Herschel Island, it was always in her mind that she would continue on to Aklavik if possible. The supplies she had bought in Vancouver had been sent to Aklavik, and there she would eventually get transportation to return home. The few residents of Herschel Island would have been hard pressed to accommodate an unexpected guest for the winter, and after making the journey from Martin Point to Herschel Island, Isobel was not in the least daunted by the prospect of more winter travel.

The day after Gus Masik's departure, Ethier arranged for a native guide to take Isobel the first twenty-five miles to Head Point for the modified price of $15. She noted that long-distance travel by dogsled was becoming more expensive than flying, with the charge being reckoned at $10 a day plus the same charge for the return journey without the passenger, and sometimes extra payment for dog food.

The temperature was −20°F when Isobel left Herschel Island on Monday morning, November 10, with dawn breaking over the Endicott Mountains. Roland, the guide, was handsomely dressed, and his eight fine dogs were wearing red saddles and pompons of scarlet wool. Constable Ethier had business at Head Point and accompanied them with his own sled. The run across the sea ice was beautiful; the sun streaming upward over the mountains turned the icebergs to fire. At 4:00 P.M. they reached a house that was used by the Mounted Police and looked after by a native and his family. Isobel was able to make herself understood using Greenlandic words, and the family was interested in her colorful Greenland clothing. The daughter had been a pupil at the mission school at Shingle Point and had a good collection of dried flowers, which she offered to share with Isobel.

A young Inuk named Isaac from the neighboring house was willing to take Isobel and her heavy luggage as far as the mission at Shingle Point. This was her first journey alone with a native guide, and conversation was limited. It was nearly twenty degrees colder than the day before, and Isobel often had to run behind the eight thin dogs just to keep warm. The picturesque trail, winding southeast between mountains, followed the bed of a river and crossed several lakes. At a brief stop to warm up at the home of Isaac's parents at King Point, Isobel shared her thermos of coffee and accepted a *mukbarak,* or bannock biscuit, the only food in the house. There were fifteen more miles to go, and it was dark and bitterly cold before the light in the mission outpost at Shingle Point appeared.

It was the hardest day of travel Isobel Hutchison would experience in the Arctic, and the intense cold was only part of the story. It was the only day with no non-native influence to provide some semblance of comfort. She had little contact with her hired driver, and the agreement did not include the well-stocked "grub box" that Gus Masik carried as a matter of course. The journey between Shingle Point and Aklavik was organized on her behalf by the Mounties or the mission at Shingle Point, and although the drivers were natives they understood the needs of non-natives.

Within half an hour of arriving at the Shingle Point mission, Isobel was seated at a table spread with a white cloth and silver cutlery enjoying an excellent meal, discussing the Oxford Group Movement and kindred topics with three English ladies and Mr. Webster, the young Anglican missionary. The residential school of thirty-eight pupils was run by Mrs. Butler; Miss Jones, a deaconess, supervised the building the girls lived in; Miss Marian Harvey was the housekeeper in charge of the kitchen where the older girls learned cooking; and Mr. Webster looked after the boys' house. They were assisted by a native deacon, Thomas Buick, who was away hunting caribou to stock the school's larder.

During Isobel's stay in the cabin on Sandspit Island, Gus Masik had frequently railed against missionary teaching that undermined the native people's ability to survive in their environment. It would be almost fifty years before this view was recognized and changes were implemented; residential schools are no longer acceptable to the aboriginal people. The mandate of the schools then was to assimilate the children into mainstream society through education in English or French. This was done at the expense of native culture and language, and children were often punished for speaking their own tongue while they were at the schools.

From her experience in Greenland, Isobel was in favor of the Eskimo receiving the "benefits" of civilization and was not aware of the destructive force such teaching could have on native life. In her travels she had a chance to observe Eskimo life under three different jurisdictions—Danish, American, and Canadian. Shingle Point, a good example of such an institution, was her first connection with a Canadian residential school and impressed her favorably.

At the Shingle Point school, most of the children came from homes in the Mackenzie delta; only one or two were from more distant points in the eastern Arctic. Isobel observed that they were taught the English language, English folk dancing, and to recite English poetry, and she realized these skills would be useless to them when they returned to their families on the trapline. Mrs. Butler told Isobel she occasionally longed for London while teaching at the only Inuit school run by the Anglican Church of Canada.

Some staff members of residential schools may have shared the sentiments of Miss Eden Corbett, who resigned her teaching position at an Anglican school in 1944 because she believed the educational process she was participating in was not just ineffective but morally questionable. The system did not meet the requirements of the natives, and segregating the children from their families was contrary to Christ's teachings.[3]

In 1925–27 an influenza epidemic had swept through the whole of the Mackenzie River valley from Fort Smith to the delta, and many of the people of Shingle Point had died. The elders and medicine men were particularly vulnerable to the sickness, and their deaths left those who survived leaderless and dependent on the traders and the church.[4] In addition, tuberculosis and typhoid were an ever-present threat. Just two weeks before Isobel arrived at Shingle Point, Nurse Tomalin and two of the pupils had been taken to the hospital at Aklavik by airplane suffering from typhoid. A further misfortune had occurred when the house Mr. Webster was living in was accidentally burned to the ground, leaving him with only the clothes he was wearing.[5]

The mission buildings on the sandspit were in constant danger from storms in the fall and from the ice that piled up close to the doors in winter. In the year following Isobel's visit the Anglican Church allocated $10,000 to replace the buildings, which were reputed to be badly overcrowded, but by 1936 the school was closed and relocated in Aklavik.[6]

Despite its troubles, the mission was a haven to Isobel while she waited for transportation onward. The staff welcomed the rare opportunity of a visitor from the outside, and Isobel enjoyed the clean, well-taught children. Her departure from Shingle Point was timed to coincide with the trip Thomas Buick would make to collect the Christmas mail, due in Aklavik about the end of November.

During Isobel's stay, which stretched to thirteen days, another visitor, Dan Crowley, arrived at the mission bringing mail to be taken to Aklavik. Crowley, an American with a wide Irish smile and natural diplomacy, was field superintendent for the Lomen Brothers of Nome. The Canadian government had contracted with Lomen Brothers to deliver three thousand reindeer to a corral prepared for them on the east side of the Mackenzie River. The drive had taken four years owing to problems with the route, weather, and stampedes in which large numbers of reindeer turned back and raced for home. The Lapp herders, facing terrible hardships, had brought the reindeer through the mountains south of Shingle Point and were preparing to take them across the sea ice of the Mackenzie delta when the ice was firm. Crowley was in the area to coordinate the last maneuver with the Canadian representative, Erling Porsild.[7] Isobel had met Porsild's father, Morten Porsild, in Greenland five years earlier, and father and son had helped her identify for publication the plants she collected.

In the early morning darkness of November 23, Isobel and Thomas departed for Aklavik, accompanied by Mr. Webster with his own dog team. After two hours on the trail, with the temperature dropping as the wind rose, the bitter cold forced them to turn back. When they set out again the next morning at 5:30, the wind had dropped. At midday the route turned south, joining the Moose River channel of the Mackenzie delta, and at sunset they found an empty shack with a good stove and shavings ready for kindling. With a storm brewing, the travelers were joined in the cabin by two men going in the opposite direction: Charlie Stewart, the father of Mrs. Ethier at Herschel Island, and Constable Mackenzie, calling at the cabin to stock it with firewood. Mackenzie, a Scot, informed Isobel that she was "wanted by the police" at Aklavik but had been given up as "a bad job." Isobel continued to think it a joke that the Mounties should be concerned about her.

The storm had blown itself out by morning, but heavy snow obliterated the trail and caused much delay as they searched for it in temperatures of thirty below zero. Thomas ran a trapline along the route, and Isobel dreaded coming upon a trap with a living fox caught by the paw. On the trail they met Jim Kane, a popular trapper in the delta whose home, a day's journey from Aklavik, was open to all travelers. As they stopped and made a campfire to thaw their frozen fingers, Kane suggested they spend the night at his trapping tent, closer than the one belonging to Thomas that they had been aiming for.

The tent was hidden in the scrub at the edge of the tree line—the first trees Isobel had seen since the Yukon River, only four months before but a lifetime earlier in experiences. The tent floor was spread with spruce twigs, and a stovepipe pushed out through the canvas roof. Isobel set up her small tent as a sort of annex, its flap open to the warm entry of the large tent, and slept well at −30°F.

As Jim Kane conducted them to his home on the bank of one of the main channels of the Mackenzie River the following day, Isobel was enchanted by the scenery. They reached his house too early to stop for the night and pushed on to the home of a young couple, Mr. and Mrs. Baker, who welcomed them. Mrs. Baker was practicing her violin when they arrived, and the walls were hung with her oil paintings. After supper she cleverly divided the single room into three compartments using string and several Hudson's Bay blankets.

Unlike the desolate area of the Alaskan coast, the route from Shingle Point passed through an area with an extensive history of contact. The Mackenzie River was a highway for the fur traders, giving access to the fur-rich Mackenzie delta to both the Hudson's Bay Company and independent traders from early in the nineteenth century. After Alexander Mackenzie's exploratory canoe voyage to the mouth of the river in 1789, fur trade and exploration went hand in hand, and the Hudson's Bay Company ruled the region like a fiefdom. In contrast to Alaska, management by commercial interests existed throughout the nineteenth century, and when the company gave up control of the territory to Canada in 1870 the foundations for government had already been laid.

Influenced by the Mackenzie River, the tree line curves north, giving an aspect of shelter to the land, and even though recorded temperatures are as low as on the barren coast, the trees moderate the ferocious Arctic gales. Abundant wood allowed for more elaborately constructed native dwellings than on the Alaskan coast, where sod, skin, and whale bone augmented the driftwood that came ashore. Rasmussen had found the Inuvialuit, the Inuit of the Mackenzie, who were influenced by their contact with Caucasians, more sophisticated and less openly hospitable than their kin farther east. On her journey up the river Isobel met many people apart from natives who had chosen to live in the North.

On a beautiful day, Isobel rode the last fifteen miles into Aklavik, comfortably wrapped in an eiderdown in a sleigh belonging to Constable Mackenzie, who had caught up with them as they were leaving the Baker home. Her goal had been achieved. Six months and thirteen thousand miles had brought her from Manchester to Aklavik.

Isobel was exhilarated and for the moment felt invincible. Having completed a difficult journey in winter conditions, she believed there was nothing she could not tackle. When she set out, her expectation was to travel the whole route by scheduled boats, including her passage south on the Mackenzie River to connect with the train that would take her to a ship on the east coast. How differently it had evolved! What experiences she had encountered and what friendships she had made!

On a high clay bank overlooking a bend of one of the main channels of the Mackenzie River, Aklavik was the most northerly point of call for the Hudson's Bay Company boat the *Distributor.* Had circumstances been different, Isobel would have boarded it for her return south. As it was, she found herself in a small native village dominated by three hierarchical institutions: the Royal Canadian Mounted Police, the Hudson's Bay Company, and the church, both Anglican and Catholic. Suddenly her accomplishments on the trail seemed to shrink in the face of such authority.

Aklavik, a mere trading post in 1917, grew to replace a large Inuit community at Kittigazuit, on the east side of the delta, decimated by an epidemic. By 1933 it had expanded to contain the Hudson's Bay Company and three other trading posts, two mission hospitals, a church, a post office, a wireless station operated by the Royal Canadian Corps of Signals, a new roadhouse, and a scattering of log houses for the Inuit and the Gwich'in, also known as Loucheux, when they came in from their traplines. The fur trade, though in the doldrums in the 1930s, was the lifeblood of the village. The white population, representing government, commerce, and the church, was mainly transient, looking ahead to the next posting—a normal feature of northern life.

The RCMP establishment was a dominant feature of the village, occupying the traditional square compound, with houses for Inspector Rivett-Carnac and Corporal Parkes and barracks for four constables and three special constables in addition to warehouses and an office building.[8] Dr. J. A. Urquhart, the government agent in charge of administration of a large area of the western Arctic, also served as chief medical officer for the area. The first airplane had landed in Aklavik in 1929, and the village was the northern terminus for Canadian Airways.

Unable to find an empty house to rent for the winter as she had done in Greenland, Isobel spent her first two nights at Mrs. Kost's roadhouse. Dr. Urquhart arranged for more suitable accommodations at the Catholic hospital and residential school, operated by the Grey Nuns and serving mainly the native community. This imposing two-story building was at the far end of the straggling village.

The social life within the white community consisted of an occasional dinner party, perhaps with a game of bridge, when the inspector or the doctor entertained the Hudson's Bay post manager along with wives and some of the unmarried nurses.[9] Richard Bonnycastle's arrival in Aklavik on inspection tours for the Hudson's Bay Company, usually in the summer, often precipitated a round of social activity. Bonnycastle's diary of June 22, 1930, commented on the prevalence of gossip within the small white community. That same summer Bonnycastle and also the Hudson's Bay post record noted the presence of two lady writers in Aklavik, and the tone of both records is cool. The request by one for accommodations on Herschel Island was peremptorily turned down.[10] The implication was that the Arctic was no place for a woman unless she was married to an official or was a nurse or teacher.

In such a tight little community, where the social structure was clearly defined, Isobel, a single white woman, was an outsider who did not belong. She had previously experienced life in an isolated northern community at Umanak in Greenland, but the atmosphere in Aklavik was different. Having received the blessing of the Danish government to work as a scientist in Greenland, she was accepted by all without question. In northern Canada single women, unless they were nurses or nuns, were viewed with suspicion by the male population. All the white men had positions within the structure and jobs with attendant responsibilities, the fortunate ones being accompanied by their wives. A botanist in a world heaped with snow had no apparent occupation, no position, and no business being there.

To some Isobel was a novelty and a breath of fresh air in that closed society. Her diary recorded visits with several women of the community: Mrs. Bonshaw, Nurse Tomalin from Shingle Point, Nurse Bradford, and Mrs. Urquhart. Mrs. MacLean, the postmistress and organist at the Anglican church, was particularly kind, inviting Isobel to dinner more than once, as did the Reverend Mr. Murray and his wife.[11] To others, such as Mr. Banks, post manager for the Hudson's Bay Company, and his wife, the independence she had shown set a dangerous precedent, and they chose to ignore her. Before the advent of the airplane, the Hudson's Bay Company controlled the transportation system and hence access to the North. Her reception in the white community varied from warm to icy, and in contrast to Greenland, she had no friend like her kivfak, Dorthe, to give her an entrée to the native community.

Always self-sufficient, Isobel took an interest in her surroundings at the hospital, worked on her new book, *North to the Rime-Ringed Sun,* and made friends with the nine nursing sisters, Father Trocellier, and the lay brothers and with some of the patients, many of whom were suffering from tuberculosis. Among the patients was an old Inuit *angakok*, or shaman, named Apakag or Apagkag. When he was first brought to the hospital by Father Trocellier suffering from heart trouble and asthma, he refused to live indoors, establishing himself in his tent in the yard, where he did his own cooking and washing. Now too frail, he occupied the room across the corridor from Isobel. A decade earlier, in 1924, Knud Rasmussen had met this same man and declared him a first-rate storyteller. Although Apakag was extremely deaf, Isobel communicated with him in a limited way with her few words of Greenlandic and occasionally played checkers with him.

December 1933 was extremely cold with many gales, duly recorded at the Hudson's Bay post, where the weather was a subject of interest when nothing more noteworthy occurred. The mail plane, expected at the end of November, did not arrive until December 19. Isobel usually attended the Anglican church halfway across the village where Mr. Murray conducted three services each Sunday, in English, Gwich'in, and Inuktitut. Isobel enjoyed the native services as much as the English, but when the weather was bad she attended mass in the chapel in the hospital.

Christmas enlivened the village, with natives arriving from the traplines, Christmas concerts in the two schools, decorations, and gifts. Inspector Rivett-Carnac and his wife, Mary, served Christmas dinner to the men under his command. The Mounties' scarlet jackets created a festive atmosphere, along with a bouquet of paper roses that had been presented to Mrs. Rivett-Carnac at the birth of their baby daughter, two days after Isobel arrived in Aklavik.[12]

Isobel was also a guest at that dinner, although Rivett-Carnac's only reference to her in his book, *Pursuit in the Wilderness,* is in a footnote, as "the lady botanist." In her diary she recorded frequent visits from Mary Rivett-Carnac, either by herself or with her husband. The casual way Isobel arrived in the settlement negated the chivalrous concerns Rivett-Carnac felt for the "unattended lady" stranded on the Arctic coast. He had made the inconvenient trip to Herschel Island after the end of the normal season for boat travel to "rescue" Isobel, at a time when he was concerned for his wife. Rivett-Carnac, born in India and educated at private schools in England, had led an adventurous life in India before coming to Canada to join the RCMP.[13] Undoubtedly he and Isobel found many points of shared interest.

Another patient at the hospital where Isobel lived was Axel Rosen, a Swedish trapper who had injured his eye. While he was recuperating, he and Isobel had many discussions about his life and travel in the North. It was normal for Isobel to be interested in the stories people had to tell, and as always she was open to ideas and suggestions. Why not? She had already proved herself capable of winter travel, and Aklavik had few charms.

Although planes flew south sporadically during the winter, there was no definite plan as to when she would leave. With Rosen she explored the feasibility of traveling "out" via the Arctic coast, Victoria Island, and King William Land to Hudson Bay—the reverse of the route Knud Rasmussen had taken, and Rosen was eager to be her guide. But when she discussed the idea with others she was warned against it because of the extreme severity of the weather and the lack of dog food on the coast.

The Mounted Police were charged with the safety of visitors to the North, and they were well aware of the dangers of winter travel. In 1911 four members of a Mounted Police patrol had died making a winter trip from Dawson to the Mackenzie River post of Fort McPherson.[14] Having observed Isobel's independence and acceptance of tough travel, however, Rivett-Carnac discussed ideas for further Arctic travel with her on at least two occasions. At Christmas dinner he said it was neither safe nor possible to go east by sled but suggested that she might go as far as Cambridge Bay, returning in the summer by boat. Four days later her diary recorded: "Inspector arrived with new idea that by taking the first plane out and getting connection at Resolution, I might reach Coppermine and get to Victoria Island. All agog and ready to try for North-East passage. Getting on from there to Chesterfield via King William Land." Calm reason returned the following day: "Pilots advise against it—so decided not to try N.E. passage."[15]

In all of her adventurous travel, Isobel had acted on suggestion. Having reached Aklavik against all odds, she was undaunted by the risk of cold and hardship and felt strong enough to undertake anything. Always in her mind was the example of Rasmussen and the young Inuit woman who traveled with him. But one does question her understanding of the geography. When she reached Chesterfield Inlet, how did she expect to proceed?

The opportunity to fly north to Coppermine was too much to resist. She must first fly south with the mail plane to Fort Resolution on the south side of Great Slave Lake to connect with the airplane flying to Great Bear Lake and Coppermine. On short notice, Isobel packed and was ready to leave on January 1, but owing to the excessive cold, −56°F, the flight was postponed. The following day, all farewells said, Isobel was settled into the tail of the four-seater Bellanca, amid the sacks of mail and furs, with only a square inch of unfrosted glass for a view. The airplane refused to start, and she was sent back to the hospital to wait. By midafternoon she learned that the engine had been badly damaged by the low temperatures and a new one would have to be sent from the south.

Sitting among her unpacked bags in the hospital room, Isobel Hutchison wrote a long letter to H. H. Rowatt, chairman of the Dominion Lands Board, giving what the officials later considered "a meagre account of the difficulties encountered while en route from Alaska to Aklavik."[16] The opening lines of her letter show that it was prompted by a conversation with Inspector Rivett-Carnac that took place after Isobel left the stricken plane and returned to the Grey Nuns' residence: "A copy of the letter which was sent to me by you, to Aklavik on 7th July 1933, has just been shown to me by Inspector Carnack [*sic*] RCMP."[17] The package sent to Aklavik, containing the original of the letter, the official permit, information booklet, and maps, had been returned to Ottawa marked "not known," and it was assumed there that she had not carried out her original plan.[18] It may seem strange that Inspector Rivett-Carnac had just given Isobel the copy of the returned letter, but in fact it had arrived from Ottawa only the previous day. Although flight had revolutionized communication between Aklavik and the outside world for most of the year, in winter the community was nearly as isolated as before the time of the airplane.

Isobel went on to outline her projected plans:

> I am very much obliged to your kind permission to collect ethnographical material on Herschel or in the Mackenzie delta, though I fear that now I may have to return by airmail in early spring (February) before any digging is possible. Should I by any chance find I can remain till the summer boat, I shall let you know, and would be very grateful for the map, book and permit. It is possible if I go to Scotland in February, I may be able to return next year and carry on the work which the exceptional ice conditions round Alaska in 1933 prevented as far as Canada was implied. I concluded my collecting for Cambridge Museum therefore in Alaska at Barrow and forwarded some interesting material from there, also plants for Kew. I should be much interested in returning to the Delta or Eastern Arctic if possible and I know Cambridge would have been much pleased had I been able to reach Cambridge Bay on Victoria Island. I contemplated the journey there by plane to Coppermine and thence by dog-sled this month and was ready to start yesterday when the plane broke down, owing to the extreme cold this year. So I fear I must leave it for a better opportunity.[19]

As always, Isobel was trying to keep every possible option open. Her emphasis on collecting for Cambridge and Kew would carry considerable weight with government officials, although it is hard to understand why it would matter to Cambridge University if she reached Cambridge Bay. As she described the projected journey there, aborted by the severe cold, she treats it as nothing out of the ordinary. Unfortunately we do not know the reactions of the officials reading her letter in Ottawa. They would find it hard to accept that anyone, especially a woman, could be so casual about travel in "their" North at any season.

Rivett-Carnac had reported Isobel's arrival in Aklavik to Ottawa on November 27, 1933, and also that she would be leaving on the mail plane in February.[20] (This letter would not have reached Ottawa until late in December because of the weather.) In truth her plans had been in a state of flux ever since she arrived in Aklavik. Having made such effort at considerable expense, she did not want to miss any worthwhile opportunity.[21] Early in December the Reverend Mr. Murray had "advised" her to visit Coppermine, although Inspector Rivett-Carnac told her it would be a tough trip, taking more than a month. On a visit to Mary Rivett-Carnac on December 21 she "promised" to fly out in February.[22] It must have been a relief to the inspector to have her promise repeated after the attempt to fly out in January failed. He, after all, was charged with the safety of this charming and intelligent, but impulsive and unpredictable, woman.

The acting chairman of the Dominion Lands Board replied to her January letter that the permit, maps, and regulations would be mailed to Scotland and asked her to advise them of her plans if she decided to return the following year.[23] There is no record of her applying for permission to return to the Canadian Arctic.

With her plan to visit the eastern Arctic frustrated, Isobel was not discouraged from further travel in the North. On a trip to Aklavik early in December to collect the mail, Erling Porsild invited Isobel to visit the reindeer station, where the arrival of the herd was imminent. The station was sixty miles north of Aklavik near the abandoned community of Kittigazuit, on the eastern side of the Mackenzie delta.

With Axel Rosen and his dog team, Isobel set out on January 10 in mild weather, but with heavy snow making their progress so slow that they were prevented from reaching the halfway house. They spent the night in a trapper's tent with an Inuit couple, Martin and Laura, and their baby, Agnes, also held up by the heavy snow. The tent was old and ragged, they were damp from the day's travel, and the night was bitterly cold in spite of the stove. The next day they reached the halfway house, a large log cabin owned by a generous Inuit couple, the Simons, whom Isobel had met in Aklavik. The Simons kept open house on a heavily traveled route, and including the family there were sixteen sleeping in the large single room that night. One of the guests was Dan Crowley, the American Isobel had met at Shingle Point on his way to the reindeer station. One more day of travel down the broad east channel of the Mackenzie River at −42°F brought them to the Porsilds' comfortable log house.

Erling Porsild, who had grown up at his father's research station on Disko Island in Greenland, just south of Umanak where Isobel had wintered in 1928, had lived most of his life in Greenland, spoke fluent Greenlandic, could handle a dog team, and had all the skills needed for Arctic life and travel. He had a doctorate in botany from the University of Copenhagen and was doing postgraduate work at the University of Chicago in 1926 when the Canadian government hired him and his brother, Bob, to re-search the feasibility of introducing a reindeer herd into northern Canada. Together they trekked the fifteen-hundred-mile route the reindeer would follow to assess the quality of the grazing, and then he explored the whole Mackenzie delta area for a site to receive the reindeer. Bob had left the project in the summer of 1933, and Erling and his wife, along with several Lapp couples hired as herders, remained at the reindeer station eagerly awaiting the arrival of the herd.[24]

Three days after Isobel and Axel Rosen arrived at the Porsild home, Erling, Dan Crowley, and Rosen left for Kittigazuit, where the corrals had been built to receive the reindeer. Isobel spent two happy weeks with Mrs. Porsild, skiing on trails cut through the willows bordering the hills behind the camp. From the summit of the Caribou Hills there was a wonderful view across the delta at sunset. Mrs. Porsild was Danish, and Isobel was pleased to use the language again after five years. They found many topics of common interest from the wider world of Europe and Greenland, a pleasant change for Isobel from the narrow, parochial life of Aklavik.

Isobel hoped to see and photograph the reindeer herd, but at the end of two weeks Axel Rosen arrived back at the camp with the disappointing news that once again the animals had spooked in a blizzard when they were halfway across the delta to Richards Island, stampeded back to the area they had come from, and disappeared south of Shingle Point. It would be another year before the reindeer drive was successfully completed in the spring of 1935.

Another mail plane was expected in Aklavik early in February on which Isobel was to fly out, so the return from the reindeer camp was made in two days. With stops to warm up at other houses along the way, they reached the Simons' house and stayed overnight, again with Martin, Laura, and baby Agnes, their travel companions at each stop on the trip out. It began to snow as they set out in the morning, and the trail became progressively worse. Isobel and Axel had to walk most of the way behind the exhausted dogs. Martin, Laura, and their screaming baby were following, planning to spend the night in the trapper's ragged tent. This gave Axel the incentive to push forward the final fifteen miles to Aklavik. At 8:00 P.M. they stopped, built a fire, and rested for a couple of hours before tackling the last five miles of the trail. Isobel knocked on the hospital door at 2:00 A.M. and was led down the corridor to her old room. Bed was never more welcome. They had been sixteen hours on the trail.

Within a few days the plane arrived. Flying in the North in winter was in its infancy. Extreme cold, long hours of darkness, landings on drifted snow, and sudden blizzards were just some of the hazards that made the fifteen-hundred-mile flight from Aklavik to Fort McMurray as much an adventure as everything that preceded it.

"On a dark morning, the 5th day of February, just as the first red streak of dawn was stencilling the inky Rockies, I looked my last upon Aklavik."[25] After stops for mail, furs, and fuel at Fort McPherson, Arctic Red River, and Fort Good Hope, the four-seater Bellanca piloted by Mr. McMullen landed before dark at Fort Norman. The "waiting room," a mile from the village, was a tent with coffee brewing on a stove. Mr. and Mrs. Douglas, friends of the pilot, insisted that Isobel too stay the night with them.

A snowstorm delayed takeoff until noon, but they were able to reach Fort Simpson, where Isobel stayed with the Grey Nuns, of the same order as her friends in Aklavik. The weather was kind, but a forced landing on a lake near Fort Simpson to adjust a minor problem caused delay. After Fort Simpson, there were stops at Fort Providence, Hay River, and Fort Resolution, where Isobel was allowed time for a cup of tea with Mr. Mason, a Scot who had left Edinburgh in 1882. It was another short hop to Fort Smith, and following the Slave River they reached Fort Chipewyan just before dark, with the wind rising. They were welcomed at the trading post by Mr. and Mrs. Reid, and Isobel shared a bed with a young nurse from Edmonton.

A howling gale gave them a bumpy two-hour flight to their destination, Fort McMurray, and once again Isobel was given hospitality, this time by Walter Gilbert, the chief pilot for Canadian Airways, and his wife. In three and a half days Isobel Hutchison had traveled the route first seen by Alexander Mackenzie in 1789 and followed later by fur traders, prospectors heading for the gold rush, missionaries, surveyors, and many others going to the North.

In the morning Isobel took a taxi to Waterways and connected with the weekly train to Edmonton. She continued by train through Winnipeg to New York, where she boarded the liner *Alaunia* and arrived in England about the beginning of March. She had been away nearly a year, and the range of her travel was exceptional—from the Panama Canal to the Beaufort Sea, using transportation ranging from ocean liner to dog team.

News of Isobel's adventures preceded her home. John Bythell, a pilot with Canadian Airways returning from Aklavik to Edmonton early in January, was so impressed with her story that he gave it to the press. Isobel was greatly surprised to be pursued by reporters from Edmonton to Boston.[26] A concise three-paragraph item of understated accuracy appeared in the *Times* of London on January 10, the day Isobel was setting off for her visit to the reindeer camp. The following day the London *News Chronicle* increased the coverage and gave the impression that she was traveling completely alone with the statement, "Miss Hutchison hired an Eskimo dog team and set off by sledge on a 500-mile journey along the frozen coast to Herschel Island." Reuter had the story from "Port Murray," Alberta, and inflated it further: "She had hoped to connect with the Arctic boat Pattason [*sic*] at Point Barrow, but failed. She therefore set out on her journey eastward on Eskimo schooners which she chartered from the tribesmen." The London *Daily Mail,* with the dateline Winnipeg, February 3, building on the story by John Bythell, had Isobel not only chartering an Eskimo vessel but setting out alone to complete the trip with borrowed dog teams![27]

Coverage became accurate and hyperbole disappeared when reporters interviewed Isobel in person. Reporters from the *New York Times* and the

New York Sun pursued her to the boat in Halifax, where the *Alaunia* called on its passage from New York. In their headlines on February 27, both made much of the fact that she had caught her first cold of the whole trip in New York, which according to the *Sun* she blamed on the dampness. In a letter to Stefansson, who had sent her the clipping, Isobel heatedly refuted that comment, unashamedly blaming the cold on "the shocking overheating in the American trains."[28] In the *New York Times* interview she announced her intention of returning to the Canadian Arctic to search for plants. A long article in the *Edinburgh Evening News,* March 13, written after an interview with Isobel at home in Carlowrie, dealt almost entirely with the plants and artifacts she had collected, the travel story having been carried by the same paper four days earlier. It ended with a prediction that she would be setting out for the North within the year.[29]

Plant collecting in Greenland had been important to her, but collecting for Kew on this journey had moved Isobel a notch higher on the scale of amateur botanists. Beginning at the Scientific Congress in Vancouver, she had met serious professionals and committed amateurs both in the Yukon and at Nome, and she was developing a network of knowledgeable associates.

In interviews with the press, Isobel Hutchison was careful not to mention the part played by Gus Masik. She was conscious of the opprobrium society would attach to her having spent two months alone with him in his one-room cabin and wisely preferred to tell the story her own way rather than risk misinterpretation by journalists. This led to the exaggeration in the accounts that appeared. The two slight boasts she made were that she was the first non-native woman to board the derelict ship *Baychimo* and that she was the first one to cross into Canada at Demarcation Point.

Throughout *North to the Rime-Ringed Sun,* published before the end of 1934, Isobel Hutchison was discreet in describing places where she stayed. Her experience of life in Greenland had prepared her to accept whatever living conditions were available. With regard to persons, Isobel was equally circumspect. She expressed warmth for those who became her friends but was rarely negative in her comments about others. She identified best with strong, competent persons, including those living rough, ordinary lives, such as Ira Rank, the Palsson brothers, and Gus Masik. Fluency in the native languages would have helped her enjoy greater contact with the Inuit and Gwich'in people. The white society of Aklavik, with its small talk and gossip, was of little interest to her and is rarely mentioned.

Isobel carried a considerable amount of money from the time she left Nome. She bought many artifacts along the north coast of Alaska, and in Barrow she cashed a check for $300 to pay for board and lodging with Gus Masik and for the trip by dog team to Herschel Island and on to Aklavik. Earlier travelers, Lady Clara Vyvyan and her friend Gwen Dorrien Smith, had crossed the continental divide from the Mackenzie River in 1926 with Gwich'in guides, carrying $500 each in cloth bags worn around their necks to pay the guides.[30] Carrying so much cash in deep wilderness made the two women feel nervous and vulnerable, but if Isobel Hutchison had such feelings she kept them to herself. With her faith in people and her complete trust in God, she slept easily.

In mere miles Isobel had traveled a long way, but in terms of cultural experience she had traveled even further. "But I had heard the call of the wild on star-lit nights under the Northern Lights; I had slept in a snow-hut; I had broken a new trail at the foot of the splintered Endicotts, and my heart beat for the wilderness."[31] The shy, self-conscious novelist of ten years earlier had become the confident "hard traveller" who had tackled difficult terrain in the depth of winter and was eager for more. She had met the media on her return to the "outside" and faced down the most prying reporters without flinching. The North was in her blood, and she would return.

II

The Lure of Distant Horizons

The woman who arrived back in Britain from the depths of the northern winter was transformed from the one who had begun her travels in the Hebrides ten years earlier. On the outside she was the same polite, considerate, modest person, but inside she had a confidence and sureness of purpose that had grown stronger with each venture. By the spring of 1934 she had completed a journey that few women—or men—would have dared to contemplate. She had traveled behind dog teams and on trading boats, through swirling snow and frigid temperatures. Had she not been stopped by mechanical failure of an airplane and by dwindling funds, Isobel was ready to continue to the distant horizon, and no goal seemed impossible. The northern winter held no terrors for her. On the contrary, in the cold North she had found warmth and friendship, especially in Alaska.

Blackie and Son of Glasgow, her publishers, were awaiting the manuscript of her new book, *North to the Rime-Ringed Sun*, most of which she had already written. As with her Greenland book, Isobel wished to include an appendix containing a complete list of the plants she had collected, in their correct families, using the proper Latin names to add scientific value to her book. From Nome, and on the last boat leaving Barrow, she had posted boxes of material to Kew. When she reached Winnipeg, Manitoba, on her journey home she mailed the last two hundred specimens of her collection. After the ship docked at Southampton in March, she visited the Botanic Gardens at Kew, and they promised the list of specimens would be compiled. With the publisher pressing a deadline of June, Kew produced the list for her toward the end of the month.[1]

Kew then turned its attention to payment for the plants. They had received specimens of 308 species, all of them considered "very good," and had promised to pay at the rate of £2 10s. per hundred. With a flourish of generosity, they forwarded to Isobel a check for 10 guineas (£10 10s.), which represented an additional £1 per hundred.[2] The value of the British pound was relatively low in 1934, and this represented less than $40. Considering the expenses Isobel encountered on her expedition, including the cost of postage, plant hunting was not a lucrative occupation. Her response to Kew was a model of grace.

> Very many thanks for your kind letter of this morning regarding payment for the plants from Alaska. It is very generous of you to give me so much and I thank you very much. I am also very glad to hear I got 308 species, I had not yet counted up exactly and thought there would be only about 250. The fun of collecting of course is not the payment but the joy of getting the flowers, and I hope I may be able to go again for Kew some day when I have saved up some more funds (if I ever do!) from my book and lectures. It is a real adventure to see a new flower.[3]

Collecting plants for Kew had a prestige value far outweighing the monetary reward. Isobel was proud of the brief account of her journey in *Maga,* the in-house newsletter of the Botanic Gardens.[4] Next came a request that she write a short account of the botanical aspect of her journey for the *Kew Bulletin,* Kew's official publication. Although they could not pay her for such an article, they felt it would have real scientific value.[5] She was startled to hear that the article should be eight to ten thousand words.

> I hope soon to get on to the article you mention for the Kew Bulletin, though I do not know if my knowledge will extend to eight or ten thousand words, even if my descriptive powers might! But I am sure it is rather the knowledge of science that the Kew Bulletin requires. However I shall be glad to try what I can and if it does not do for you it may do for *Flora and Silva* or another paper. It may be some weeks before I get on to it, as I go on holiday for three weeks on 25th August.[6]

The article was published in the *Kew Bulletin* in December 1934, and she sent copies of it to botanists she had met on her trip—Charles Thornton in Nome and Dr. Hutchinson in Vancouver. "If you can really spare another two copies, I will send one to Mr. Porsild of Aklavik, and the other . . . perhaps for Mr. Porsild's father of Greenland when the summer ship goes up to Disko Island, as he too has been writing to me about Alaskan plants."[7]

Considering that Isobel had worked feverishly since her return to Britain at the beginning of March, a holiday was justified. Without photocopying machines, she had typed the entire contents of all her botanical notebooks in order to have her own copy of them before the originals were returned to Kew.[8] The publisher received the final proofs of her book by the beginning of August. She had labored over the exact nomenclature of the plants, studying their arrangements in the correct genera and incorporating several sets of revisions from Kew. In spite of the detailed work, she also found time to write an article for the July issue of *Blackwood's*. When he received her article, Stefansson wrote, "This has been a delightful half-hour reading "Around Arctic Alaska"—Blackwoods, which you so kindly sent. Now I await the book with keener anticipation."[9]

Stefansson had previously written from New York: "When I read the enclosed article about you in 'The Sun' I was considerably miffed because you had not let me know you were in New York. It was then an embarrassing thing to find your letter of February 19 which, unopened, had got in with a pile of miscellaneous correspondence. I did so want to see you. If you are ever this way again, will you please remember one thing about me: that if I don't reply it doesn't mean lack of interest, but that for some reason a letter has escaped me."[10]

The appendix to the book also included a list of seventy-nine artifacts Isobel had purchased for Cambridge, with details of their use and the locations where they were obtained. These gave her less concern than the lists of plants, but Louis Clarke, curator of the Museum of Archaeology, was ecstatic: "The things you got for me were *wonderful* acquisitions. I am *most* grateful and I look forward to thanking you personally."[11]

In spite of her involvement in getting the book ready for publication, she had time for a thoughtful gesture. Knowing the attachment of a sea captain for his old ship, she sought the address of Captain Cornwell, the retired master of the *Baychimo,* and sent him the snapshots she had taken on her visit to the ghost ship. His letter of thanks was preserved along with the other correspondence from her new friends from the North.[12]

In July Isobel received the first of several honors that would come to her over the years. Appropriately, it was from the Royal Scottish Geographical Society, whose Mungo Park Medal recognized her researches in Iceland, Greenland, and Arctic Alaska. She wrote to Stefansson: "I certainly never expected that my feminine researches would gain a medal!"[13] Her use of the word "feminine" is significant: Isobel Hutchison had no pretensions about belonging to the exclusive class of northern explorers. Her journey to Aklavik was as noteworthy as any undertaken by male university groups of the time, yet she did not consider herself their equal. The Arctic Club, founded in 1932, elected its members from *men* who had been members of expeditions from Oxford or Cambridge, and they dined together once a year at either university.[14] Isobel had no illusions of being eligible for membership.

North to the Rime-Ringed Sun arrived in the bookstores in October, and the *Times Literary Supplement* gave it a full and enthusiastic review. "Parts of the narrative are as bare and reticent as any within the Arctic tradition; but in it there is a Kiplingesque mingling of verse and prose and a delicate awareness of detail which is the work of a naturalist who is also a woman." The journals of the two British geographical societies gave favorable reviews, the *Scottish Geographical Magazine* going so far as to say, "Her journey is, as far as we can recollect, not to be compared with that of any other woman of modern times."[15] It was later chosen as the book of the month for the Scientific Book Club of New York.[16]

The reviews made the distinction that this was Arctic travel writing by a *woman*. Not only had she told the story of an unusually daring adventure, but the fact that a woman had undertaken it added to the interest. In contrast to the grim tales of hardship that characterized much of the writing of male Arctic and Antarctic explorers, Isobel Hutchison's book emphasized the beauty of the North and the lives of those who live and travel there. She made it seem as easy and natural as a stroll through the Hebrides.

While the reviews were gratifying, the reactions of her friends pleased Isobel immensely. Her correspondence with Sir Arthur Hill, the director of Kew, which had begun with deference and formality, became increasingly an exchange of letters between friends and fellow travelers. "I am very glad you enjoyed [my book] and am delighted with what you thought of my fellow travellers and guides—they were truly fine fellows all of them, as you say—so many of those one meets in out of the way places *are*."[17] Her friends on the coast of Alaska were enthusiastic about her portrayal of their area and their lives, and Ira Rank ordered ten copies of the book to be sent to Nome. From acquaintances at the Manitoba Museum and the

University of Minnesota came invitations to visit and lecture.[18] After writing a review in *Pacific Affairs,* December 1935, Edith Tyrrell, a former resident of the Yukon and wife of J. B. Tyrrell, a redoubtable northern traveler of the previous generation, wrote to Isobel directly: "It was a big and brave thing you did."[19]

After her return to Scotland, Isobel corresponded regularly with her companions from the *Trader,* Pete Palsson and Ira Rank. Rank outlined his summer plans for 1935 to go to the King and Diomede Islands, suggesting she would find it an interesting trip and it would not be rushed. "Try and come by all means and we will do our share in every way. We will be very reasonable in our charges but we cannot promise there will not be other passengers. [Where could he hope to fit them in?] Can't promise to take you to Siberia—the officials are not friendly to American traders." After Isobel had given a fifteen-minute talk on the British Broadcasting Corporation, she received an excited letter from Ira Rank to let her know that it had been heard in Nome.[20]

In September 1934 Nome was ravaged by a serious fire that destroyed 90 percent of the buildings in the town. Isobel received firsthand accounts of the devastation from her friends there, among them Charles Thornton, the enthusiastic amateur botanist who had accompanied her on her explorations around the area. She arranged for Kew to send him a supply of proper drying papers to replace those he had lost in the fire. He continued to send seeds and plants to Isobel for Kew and hoped she would return to Nome and botanize the Sawtooth Mountains with him.[21] Even Gus Masik was caught up in Isobel's enthusiasm and sent roots of plants from Martin Point, which she was able to revive.[22]

Early in December, after the publication of *North to the Rime-Ringed Sun,* Isobel and her sister Hilda leased Carlowrie to tenants for eighteen months and moved to Newbury Lodge, Haddington, in East Lothian. During the spring, Blackie and Son published *Arctic Nights' Entertainments,* Isobel's transcription of Gus Masik's stories dictated in the cabin at Martin Point. The *Scottish Geographical Magazine* felt that "Gus had been most fortunate in his biographer."[23] Stefansson, who knew Gus, "would have liked more philosophizing and explaining, a few more conclusions. Still it is to your credit that the book is just what it is—you have transmitted the material faithfully just as you received it, thereby portraying Masik and his type better, even as you portray the country less." He also sent her a list of corrections to the text that he considered trivial but thought would interest her.[24] Mrs. Greist, in a letter thanking Isobel for the many boxes of clothing she had collected and sent to the Presbyterian Mission in Barrow, praised Isobel's books and reported that Gus was very proud of being her host.[25]

After the publication of this latest book Isobel had time for a holiday in the Scilly Isles in April before she left in mid-May to spend the summer of 1935 botanizing in Greenland. Her return to Greenland was like a homecoming with the renewal of old friendships and the discovery of new ones. On the ship from Copenhagen she met Professor William Thalbitzer, a Danish ethnologist whose work in Greenland began at the turn of the century when he wintered at Angmagssalik with his wife, an artist. Isobel, based at Egedesminde, due south of Disko Island, formed a lasting bond with Thalbitzer, who was working in the same area. Reg Orcutt, a friend of both hers and Stefansson's, arrived to make a color film of western Greenland. Filmmaking had taken him across the Far North from Ellesmere Island to Spitsbergen.[26]

Isobel saw many of her old friends as she explored the coastline around Egedesminde, including Peter Rosing, Knud Rasmussen's widow, and Dr. Gudrun Christensen, her hostess and rival in 1928. However, her greatest joy came with her return to Umanak. She brought with her the gift of a bell for the new stone church, built since her year in the village. For the dedication service, Isobel sat in the choir with her former kivfak and friend, Dorthe, and they drank many cups of coffee together during her week in Umanak. As always the returning ship called at many ports along the Greenland coast, giving Isobel another look at familiar places. One of the stops was at Godhavn, where Dr. Porsild took her on a hunt for plants.[27] In four months she collected 1,700 specimens for the British Museum, which had given her a list of plants they wanted in quantity.[28]

Already Isobel's mind was on further travel. Not only was Ira Rank suggesting interesting options, but Stefansson, on the strength of a postcard from Sir Hubert Wilkins in Russia, had ideas for her. "That's where you better go next. A good first trip would be on the two rivers Yenisei and Ob. Descend one to the mouth, round the promontory, and ascend the other. Just look at the map!"[29] This offhand suggestion shows how little

Stefansson understood Isobel's style of travel. Her preference was always for islands or seacoast, reached by ship, and quite different from the central Siberian lowlands. But when Isobel explored the possibilities of making plant collections for Kew, Siberia was at the top of her list. "I am hoping to get somewhere like East Cape, Siberia or Saghalien region [the island of Sakhalin in the Sea of Okhotsk] next year, but have nothing planned yet. Do you think Kew would like a collection from anywhere in that region if I can get it fixed up? Or the Aleutian Isles sound alluring too!" Kew had recently purchased a large collection of named Aleutian plants and was interested in what she could do in Siberia.[30]

As usual, Isobel's plans were fluid and her spirit was optimistic.

I am very sorry I am too late with the Aleutian Isles for Kew, but delighted that East Cape, Siberia, may be still of interest. I will certainly let you know as soon as ever I fix things up, if I can in the unsettled state of the world at present. However I think it is best just to carry on and take no notice and maybe there will be no conflagration after all! I am thinking of going out this time across the Trans-Siberian Railway to East Cape and perhaps coming home by the Bering Sea, but I have made no plans yet.

The long-sought trip to Siberia was not to be, and in the spring of 1936 her plans became clear.

I have not been able to fix the East Cape visit this year but perhaps another year I may be able to do so. The Russian authorities took so long to reply that was what made it too late really! [She had already made application to Russia for the following year.] I have however now decided to visit the Aleutian Islands in June.[31]

Although the Hutchison sisters had private incomes and lived in a small castle staffed with servants, they did not have unlimited resources. Their incomes depended on the trust set up by their father at the turn of the century, and the value was gradually being eroded. Without support from Kew, Isobel cast about for another sponsor. Having worked for the British Museum in Greenland, she approached them with her plans and gingerly broached the subject of payment. "Would the British Museum be likely to purchase the collection this time, as I am sorry I might have to ask this [for] this year owing to the expense of this journey." The reply was entirely satisfactory; the museum was prepared to make a grant toward her expenses rather than paying for plants as Kew had done. Later she wrote:

> I have been going into the question of the expenses of the Aleutian Islands trip, and fear it may be in the nature of some three hundred pounds for a six month's journey returning by Japan if possible and perhaps being able to collect plants from the Kamchatkan region also, at least if the season when I reach that part permits. . . . My own funds available for the six months are just about two hundred pounds, but I do not know if the Trustees would be willing to make a grant of as much as the other one hundred pounds?[32]

Without waiting to hear if her request had been approved, Isobel booked her passage to Montreal on the *Andania,* sailing from Scotland in the last week of May.

The urge to see distant lands overrode all practical considerations. Released for the time being from the responsibility of the estate of Carlowrie, Isobel could spread her wings and fly to the ends of the earth. At home she was a proper Edinburgh lady, involved in the cultural and political life of the city and in local church events, visiting sick and infirm elderly relatives, and fitting all of that around constant writing for journals, newspapers, and magazines as well as lectures and radio broadcasts. When she headed for distant horizons aboard ship, her life acquired space and freedom.

"Why must the heart always desire the inaccessible?" These were Isobel's words as she set off in 1936 for the Aleutian Islands. Her heart's desire was Attu Island, the most westerly of the archipelago and the most difficult to reach. She had read the report of the Swedish botanist Dr. Eric Hultén, who visited the islands in 1932 and reported species of plants on Attu not found on the other islands, giving them a possible link to Asia.

Treeless, windswept, and foggy, the volcanic Aleutian Islands, with their dark and tragic history, are the least known part of the United States, visited by commercial fishermen, whalers, and scientists but seldom if ever by tourists. From the tip of the Alaska Peninsula, the thousand-mile chain of rocks, reefs, and islands forms a barrier beside the deep Aleutian Trench of the Pacific Ocean, reaching almost to the Komandorski Islands and the Kamchatka Peninsula of Russia. Above this barrier the rising warm air of the Japanese current meets the heavier Arctic air of the Bering Sea, creating perfect conditions for the formation of storms.

The strong tidal currents between the islands cause upwellings of rich nutrients from the ocean depths that feed a diverse and abundant marine life in the sea surrounding the islands, supporting large populations of birds and of sea mammals such as sea otters, seals, and whales. For thousands of years the bounty of nature sustained the Aleut people, who are believed to have come across the Bering land bridge and turned southwest, splitting off from the Eskimo or Inuit and forming a distinct group with their own language and customs. The people moved along the Alaska Peninsula, gradually spreading outward until all the islands were populated. In the temperate climate, the population, achieving greater longevity and lower infant mortality than their Eskimo cousins in Arctic Alaska, may have been as high as 16,000 at the time of their discovery by the Russians in 1741.[33]

The riches of the sea and the skill of the Aleut hunters became the source of their tragedy. By the end of the eighteenth century, the Aleut people were first enslaved and then slaughtered by the *promyshlenniki*, or Russian fur hunters, their numbers reduced to one-tenth of the original population. The sea otter, desired for its lustrous pelt and the cause of this dark period of Aleutian history, was hunted almost to extinction. Word of the grim situation traveled slowly back to officials in Russia. Administrators and missionaries were sent out to control affairs in the distant colony, bringing order and decency to the remaining Aleuts where piracy had previously reigned and at the same time imprinting the people with the Russian language and culture.

In the second half of the nineteenth century, a new scourge arrived with the New England whaling fleet, which had turned its attention to the North Pacific. The whalers, with their unruly crews, usually stopped at Unalaska in the Aleutian Islands, gateway to the Bering Sea and beyond, bringing liquor to trade with the natives and sometimes allowing their seamen to go ashore. The natives who survived the first onslaught of the Russians were now debauched by the Americans.[34]

The promise of wealth from the sea continues to draw fortune hunters to the Aleutians even in recent times. For a few years beginning in 1978, an explosion of the crab population caused a huge expansion of the crab-fishing fleet working out of the island ports until the resource was quickly depleted. Once again the men who spent days and weeks working in dangerous conditions on the stormy seas sought relief from their tensions in the bars and brothels that sprang up in these ports.[35]

Profusion of a different kind attracted Isobel Hutchison to the Aleutian Islands. Though considered by most people to be barren and desolate, their meadows are aflame with color during the short growing season from July to early September. The plant life is classified as subarctic or low arctic and would be different from the material she had collected in the north of Alaska.[36]

Landing from the ship in Quebec City, with its crowded Roman Catholic churches, black-robed nuns, and voluble French-speaking taxi drivers, Isobel felt she had stepped into an old French seaport. At the Legislature, "Votes for Women" had just been defeated for the eleventh time by a handsome majority, and the door to the visitors' gallery was slammed in her face!

Before crossing the Atlantic, Isobel knew she would be delayed by a week in reaching Alaska because all transportation was booked by cannery workers. This allowed time for a leisurely trip across Canada renewing old friendships. In Montreal she visited the Allards from Dawson in the Yukon, and in Ottawa she saw Mary Rivett-Carnac. Erling Porsild, from the reindeer station north of Aklavik, now the curator of the Dominion Herbarium in Ottawa, showed her a specimen of *Spiranthes romanzoffiana,* which the British Museum had particularly requested her to find. He also advised her to try to include the Pribilof Islands in her travels. Always open to suggestions, Isobel wrote to the British Museum asking them to request the required permit from Washington, to be forwarded to her in Unalaska. She continued west, stopping in Winnipeg to visit the Manitoba Museum, mindful of her assignment to purchase artifacts for the Royal Scottish Museum in Edinburgh. In Seattle she met both Mrs. Hartshorn and Mrs. Willoughby Mason from the Yukon, an indication that Isobel maintained contact with even the briefest of acquaintances.

Sailing from Seattle on the *Yukon,* north through the Inland Passage and across the Gulf of Alaska, Isobel reached Seward on June 19, 1936, and discovered that the travel office in Washington that had supplied the information regarding shipping schedules in the Aleutians had only a theoretical knowledge of the subject. The *Starr,* on which her passage to Unalaska was booked, would not arrive in Seward for another three weeks, and valuable time was passing as spring flowers faded and set seed. On the chance that it would sail to Kodiak Island sooner and that she could join the *Starr* there, she booked passage on the *Curaçao* for the following week. During the short fishing season in Alaska, space on the few coastal boats was at a premium, and bookings were essential. Isobel made the best of her enforced week in the dusty little town of Seward, terminus of the Alaska ferry, exploring the outskirts at the base of the mountains and collecting sixty specimens and as many mosquito bites. At the Jessie Lee orphanage run by the Methodist Church, she met her first Aleut natives when she was invited to show her Greenland slides, wearing her Greenland costume.

By including the Greenland material in her luggage, Isobel was keeping faith with the work of Knud Rasmussen. She tried to use her Greenlandic words to communicate with the native people wherever she went in Alaska, always hoping to find a connection that stretched across the breadth of the continent. On another level, the Greenlandic slide show was a form of entertainment, her way of making a return for hospitality received, as performing the sword dance had been in Greenland.

The *Curaçao* sailed as promised and took Isobel on the overnight trip to Kodiak, the large mountainous island south of Seward and southeast of the Alaska Peninsula. Kodiak is not one of the Aleutian Islands, but it provided a new area for botanizing. Accommodations at the recently built Sunbeam Hotel in the village of Kodiak were clean and comfortable. The colorful owner, Charlie Madsen, had gone north at the beginning of the century, married a young Eskimo girl, and been grubstaked to start his own trading enterprise between Nome and Siberia by Isobel's friend Ira Rank. Moving south a few years later, Madsen turned to trapping and trading on the Alaska Peninsula, where he made a reputation as a successful hunter of Kodiak bears.[37] In his new incarnation as hotelier, he employed the members of his growing family. His son, also a bear hunter, single-handedly carried to Isobel's room the large wooden crate, supplied by the British Museum, that it had taken a crane to move in Seattle and three men to load aboard the *Curaçao*.

Immediately north of Kodiak Island on the Alaska Peninsula is a cluster of volcanoes, including Mount Katmai, which had erupted in 1912, burying the northeast corner of Kodiak Island a foot deep in ash. The crater of Katmai is considered the deepest on earth, at three thousand feet from the high rim, and was viewed in 1934 by Father Bernard Hubbard, a geologist specializing in volcanoes. To reach Katmai his party crossed the Valley of Ten Thousand Smokes and observed that the volcanic action had dwindled to several hundred smoking fumaroles since his first visit in 1928.[38] Isobel, arriving twenty-four years after the first eruption, noticed that ash from the volcano could still be found in depressions and hollows, but despite this the plant life had recovered and was growing luxuriantly. During the two weeks she spent on Kodiak awaiting the *Starr,* Isobel's collection of plants and acquaintances grew.

Many of the local people were interested in her occupation and were helpful in directing her to the richest areas for plants. An Irishman named Coffin described a rare plant that grew on the Alaskan mainland, on which every petal was a different color and every leaf a different shape. Before she left Kodiak a specimen was delivered to her, and she named it appropriately *Coffiana falsifera inebriatum.*

The bars and poolroom were popular spots in Kodiak village and were especially busy on the eve and day of July 4, with firecrackers and guns adding to the noise of the Independence Day celebration. Isobel escaped to a quiet wooded shore and swam in the icy, crystal blue Pacific waters. With her penchant for swimming in frigid water, it is not surprising that her preference for travel was in the North.

As predicted, the five-hundred-ton *Starr* arrived in Kodiak on July 10 on its monthly 3,000-knot voyage from Seward to Unalaska, during which it called at sixty-five ports with mail and cargo. The *Starr,* in service since 1919 when it replaced the *Dora,* affectionately known as *"Dirty Dora,"* had a reputation as the most adventurous as well as the most uncomfortable passenger-carrying vessel in Alaska. Isobel found this statement no exaggeration. Each tiny cabin was furnished with three iron bunks, stacked so closely above one other that for a person of more than average girth only the top bunk was possible. A cold-water basin, a folding stool, and space to stand between the bunk and the cabin wall completed the arrangement. Apart from a minute afterdeck, the cabins, opening into the dining room or adjoining the engine room, were the sole accommodations for passengers. The discomforts of the boat were outweighed by the kindness of the crew, from the captain to the steward.

With sixteen stops along the way to drop off mail, cargo, and passengers, it would take the *Starr* a week to sail from the port of Kodiak along the coast of the Alaska Peninsula to its destination, Unalaska. Isobel made full use of the opportunity for hurried plant collecting allowed at three longer stops, and on board the ship she reveled in the spectacular scenery and enjoyed camaraderie with crew and passengers. Even the weather cooperated. Noted for foul weather and rough seas, the Aleutians in 1936 had a record summer for fair weather and sunshine. Despite this, many of the passengers suffered from seasickness, and her words create a vivid picture of life on board.

> For shorter stages we had the occasional company of natives, usually women and children. These were invariably wretched sailors, and their loud misery (a few feet behind the diners' backs) did not conduce to appetite, though our chef did his best with grapefruit, celery, stewed plums, and such-like tit-bits. The officers of the *Starr* dined first, but there was room for three passengers in addition at their tables and I was fortunate to be one of the lucky first sitters. The others watched us hungrily (or perhaps more often with revulsion tinged with jealousy) from prone positions in their bunks behind doors left too hospitably open.[39]

In the Shumagin Islands, near the end of the Alaska Peninsula, the truism of the North's being a small world was proved once more. On Kodiak Isobel had met an acquaintance from Barrow, and at Shumagin it happened again. A schoolteacher and his wife boarded the *Starr* in the hope of making a brief visit to the bright lights of the Alaskan mainland. The teacher's wife had been Isobel's guide to the archaeological points of interest at Point Hope when the *Trader* called in 1933 on the trip to Barrow.

At False Pass, the strait separating Unimak Island from the mainland, the *Starr* reached the beginning of the Aleutian Islands. Sandbars extending from the mainland are gradually filling the narrow passage, and sometime in the future Unimak Island will be joined to the Alaska Peninsula, a land still in the making. In his frequent visits to the area between 1926 and 1934 Father Bernard Hubbard observed conspicuous changes occurring in the specific localities he explored. He concluded that the Alaska Peninsula had originally been a series of islands where the action of volcanoes, ice, wind, and sea had created land bridges between islands, as was now happening in False Pass.[40]

Unimak Pass, the strait used by ships heading into the Bering Sea, separates the islands of Unimak and Akutan. The *Starr* was thoroughly tossed about passing through those waters, and to Isobel's regret, storm clouds veiled Shishaldin, the beautiful volcano on Unimak that resembles Mount Fuji in Japan. Akutan Island is probably the windiest of the windswept

Aleutians, and Bernard Hubbard had a scientific explanation for its severe weather:

> Here undoubtedly was where the storms are born. Here is the cradle of the tempests. For storms, all you need is air masses at different temperatures, with conditions present to start them moving. On one side of us was the Pacific Ocean with the warm air over the Japanese current; on the other side was the Bering Sea with its cold air from the Arctic current. The volcanic rift of the Alaska Peninsula plus the Aleutian Islands is an 1800 mile arc of high mountains and narrow passes. Through these passes the heavy cold air of the Bering pours to meet the rising warm air of the Pacific, and around these high volcanic peaks the storm-cloud nuclei begin.[41]

Passengers were landed through the surf where possible or given the promise of another attempt on the return trip as the *Starr* negotiated the treacherous waters and reached the island of Unalaska, the second largest of the Aleutian chain. This was the end of the line for the *Starr*, which would now return to Seward. Beyond lay the scattering of islands, some no more than rocks, curving eight hundred miles west to the end of the archipelago.

The information Isobel received from Washington before setting out explained that transportation beyond Unalaska would depend on the occasional service afforded by trading vessels and tramp steamers that ply back and forth among the islands without schedule. From her previous experience in northern Alaska, she accepted this information and set off with the optimistic expectation that she could reach Attu, the westernmost island.

The large harbor at Unalaska, sheltered by the bulk of Mount Makushin, is the natural center for marine activity throughout the Aleutians, both commercial and military. In 1936 the village of Unalaska, home to 250 Aleuts, was the main focus of native life on the island. In the harbor, opposite the village, is the naval port of Dutch Harbor, at that time the headquarters of the United States Coast Guard Bering Sea Patrol. With no hotel in the village, Isobel found comfortable accommodations in the vacated Jessie Lee Home, the Methodist orphanage for Aleut and Eskimo children that had moved its operation to Seward. She shared the house with an American nurse, an expert cook, and gratefully resumed her familiar role as chief bottle washer.

The volcanic slopes of Unalaska are clothed in a thick green carpet of mosses and lichens to the thousand-foot height, and out of this rises a tall, luxuriant growth of wildflowers. Among the most striking features are the meadows aflame with color during the short growing season from July to early September. On her first day of collection, Isobel secured eighty-five different species and found the plant requested by the British Museum, *Spiranthes romanzoffiana,* a cream-colored orchid with a fragrance like almonds. In the days that followed the collection grew as she ranged over alpine valleys and picked her way through mountain gullies.

In such a small community Isobel soon became known to both natives and non-natives. As usual, she attended church on Sundays, where the service was conducted in Russian by the archimandrite who had been one of her shipmates on the *Starr.* The village chief translated the service into Aleut, a language quite unfamiliar to her. The handful of non-native residents welcomed the novelty of having a visitor to Unalaska.

In the meantime Commodore Dempwolf of the Coast Guard, aware of her desire to sail beyond the island, was working quietly on her behalf.

12

To the Edge of the Western World

Up to this point in her journey, Isobel had relied on regular transportation that kept as closely as possible to a schedule. To reach her goal of traveling to the outer limits of North America to collect plants for the British Museum, she was willing to trust Providence and accept any opportunity that came her way.

Ten days after her arrival on Unalaska, the *Penguin,* a small schooner belonging to the Bureau of Fisheries, slipped into Dutch Harbor on its way to the Pribilof Islands to pick up agents of the Foulke Fur Company. It would be returning directly to Seattle, and Isobel was welcome to go along provided that she could obtain the necessary permission from government authorities to land on the islands, and provided also that she could find her own way back. The only possibility for return transportation was on a Coast Guard cutter patrolling in the Bering Sea. Naval vessels did not carry women passengers, but Commodore Dempwolf, using the Coast Guard tradition of "providing succor to persons in need," promised she would be picked up if no other ship was available.

St. George Island, two hundred miles north of Unalaska, was a welcome sight to Isobel after she had been thoroughly tossed for a day and a half on a sea with a reputation as the stormiest in the world. The two main islands of the Pribilofs are the remnants of ancient volcanoes, but St. George is the more barren, the black basalt cliffs rising more than two hundred feet out of the water.

The first American woman to winter in the Pribilof Islands, Elizabeth Beaman, visited St. George in the summer of 1880. Against official opposition, Mrs. Beaman accompanied her husband, a government agent, on a two-year tour of duty to the Pribilofs. Her diaries and letters provide a vivid picture of her life on the islands and of the Aleut people.[1]

When Isobel visited St. George, the *Penguin* stopped only long enough to pick up the agent of the fur company and the doctor. There was no harbor, and anchorage for ships depended on the wind's being directly off-shore. A seal-hide *bidarah,* similar to the Inuit umiak, was used to transport passengers to and from the shore. Other botanists had visited the island in the past, and Isobel knew of their collections. She had three hours ashore to explore and collect whatever plants she could find before sailing on to St. Paul Island, forty miles north.

The Pribilof Islands are the breeding ground for northern fur seals, which begin arriving in the hundreds of thousands in May and "haul up" on the beaches. The bulls come first to establish the territories where they will maintain their harems when the cows arrive four to six weeks later. Next to come are the bachelors, young males under five years old, too young to breed. When the cows arrive they immediately give birth to the single seal pup each has carried for nearly a year, and then the drama of defending territory, controlling the harems, and breeding begins. In the years when seal pelts were much in demand, a controlled slaughter took place for two weeks in July, with a predetermined number of bachelors being driven inland by Aleuts to the area known as the killing fields. There the seals were clubbed and skinned, their carcasses left for the scavenging foxes. The work was brutal and physically demanding, and during the two weeks of slaughter the atmosphere on the little islands was highly charged.

The location of the seal rookeries was discovered in 1786 when Gavriil Pribilof, sailing in the area of the Aleutians, spied migrating seals and followed them north. Once numbering in the millions, the seal population was gradually decimated by pelagic hunting from the ships of the various nations sailing in Pacific waters. After the purchase of Alaska by the United States, strict controls were put in place to safeguard the seal population by stationing two government agents on the islands to monitor the harvest carried out by the fur companies.

When Isobel arrived on St. Paul Island in 1936 little had changed since Elizabeth Beaman's time. Most of the population was at the jetty to greet the arriving ship, and Isobel found comfortable quarters in Government

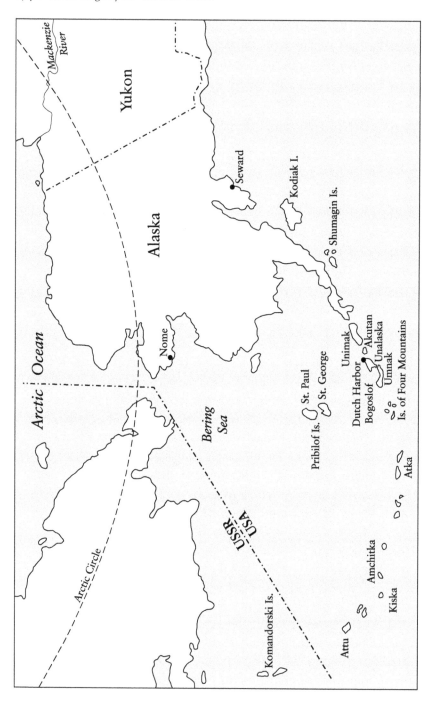

Aleutian and Pribilof Islands

House. She took her meals in the Company House, which housed the fur company employees, exactly as Elizabeth Beaman had done. What had changed was the attitude toward non-native women visiting the island. Isobel, no stranger to being alone in male company, felt at home in the clublike masculine atmosphere of the Company House.

Other new features were the fourteen miles of paved road and the automobile in which Isobel was driven out to explore, to the fullest extent possible, the small island with its beautiful sand dunes. Assisted by the American schoolteacher, she added 120 species to her plant collection on the first day, finding poppies of all kinds, muscari and musci, ranunculus, wild mustard, chickweed, lupine, painted daisies, violas and violets, anemones, and other species including tall ferns.

The timing of Isobel's visit was fortunate. Although the slaughter of seals was still evident from the piles of skinned carcasses, she was spared witnessing the event itself. As the days progressed, the Aleut men employed in the kill often resorted to kvass, locally brewed alcohol. The other residents of the island could not escape the sounds and smells during those intense two weeks.

In addition to collecting plants, Isobel spent one of the most interesting mornings of her life watching the seals and capturing their antics on film with her cine camera. "The great bulls, brownish in colour, squatted on the sands, some alone, some surrounded by their silver-coated cows, some staring out to sea, some lying asleep on the shore up which the white surf roared with a cry like that of the seals' own."[2]

On another morning she visited the Russian priest and his wife and viewed the treasures in the handsome church that replaced the original one built of driftwood by sailors in 1797. The Russian influence on the island was still strong, although the language was no longer used in the schools, which were established by the Alaskan Commercial Company in 1870 on each of the Pribilof Islands, with teachers brought from the mainland. This was more than a decade before Alaska had schools.[3]

After three days of exploring, Isobel received a radio message that the Coast Guard cutter *Chelan* would be passing close enough to call at the island on the way from Nome to Dutch Harbor, and she was requested to be ready to embark at a moment's notice. On a gray, choppy evening she was rowed out into the harbor and a swaying rope ladder was lowered over the side of the ship. Laden with flower presses, cameras, and rubber boots, Isobel gingerly clambered up and landed awkwardly on hands and knees on the immaculate deck at the feet of the watching officers and crew. Despite her ungraceful arrival on the *Chelan,* she was thrilled to have had the opportunity to visit the isolated Pribilofs.

Because of her accreditation from the British Museum, Isobel had achieved scientific standing within the small naval establishment at Unalaska. On her return to Unalaska from the Pribilofs, she met two other scientists who were excavating an ancient Aleut village at Dutch Harbor while they awaited transportation. Dr. Aleš Hrdlička, a physical anthropologist from the Smithsonian Institution in Washington, and his British assistant Alan May had spent the summer literally digging into Aleutian prehistory as far west as the island of Attu. The early Aleut people had a knowledge of embalming and were known to preserve their important dead as mummies, placing them in mortuary caves in secret locations. A burial site discovered by a sea captain in 1874 had aroused much interest and speculation concerning the subject.[4] In the course of his summer's work, Dr. Hrdlička had investigated the burial sites on the Islands of the Four Mountains, some of which were dry caves warmed by volcanic fumaroles. The locations not only were the subject of taboos but were also relatively inaccessible because of treacherous reefs in the area.

At a tea party given by Dr. Hrdlička, the boxes on which the tea cloth was spread contained mummies wrapped in their original woven grass mats, ready for shipment to the museum in Washington. At that tea party Isobel received the gratifying news that she was to be given onward passage aboard the *Chelan* when it sailed to survey the western end of the Aleutian chain.

Commodore Dempwolf had requested permission from Washington for the Coast Guard to give Isobel Hutchison every assistance in carrying out her botanical work. In addition to their regular duties of saving life and property, carrying mail, providing medical services, and enforcing the laws, particularly against poaching and smuggling, the United States Coast Guard had a duty "to provide quarters and facilities aboard ship for agents of various governmental organizations pursuing scientific investigations." This clause allowed the *Chelan* to make Isobel a temporary member of its crew of ninety for the autumn hydrographic survey cruise.

As Captain Keilhorn explained: "These seas are the most dangerous in the world. There are no harbours. We have no lights or buoys to guide us. No one can forecast the weather, since two days are hardly ever alike; the waters are full of currents of unknown force and of tide-rips. Gales and fogs are constant in summer as well as winter; in short, they could not be worse."[5] The Coast Guard had lost more than one vessel to shipwreck in those waters. Isobel would be allowed time ashore for collecting plants whenever conditions permitted, but everything depended on weather and navigational circumstances.

As if to compensate for the hazards, the accommodations on board the *Chelan* were the most luxurious Isobel had experienced in Arctic waters. In addition to a comfortable bunk and an armchair with reading lamp, there was a desk with many drawers to hold botanical paper and presses, a radiator to dry specimens on, and a commodious built-in closet in which her modest wardrobe occupied only two hangers. Despite the rough seas, she put in an appearance at all meals; the menu, in the American naval tradition, was varied, generous, and expertly served in the officers' mess. It was a far cry from the conditions she had known on the *Trader* and the *Hazel* in 1933, or even on the *Starr* earlier that summer.

The *Chelan* passed close enough to Bogoslof for Isobel to photograph the island, relatively newly created by a submarine volcano. Over the course of a few years from the beginning of the century the topographical features of this latest land formation were observed to change from a single island to two peaks connected by a long sandspit. The Coast Guard had a duty to inspect Bogoslof each year, and the intention was to visit it on the return journey, but heavy surf prevented a landing and Isobel explored the strange island only through binoculars. Continuing west, they passed Umnak Island and the Islands of the Four Mountains, the home of the burial caves containing the mummies, before reaching their first stop, the island of Atka.

Excavations show the existence of Aleut settlements on Atka reaching back over two thousand years when the people lived in *barbaras,* underground communal or single-family dwellings built of driftwood and whale bones with sod roofs. As on the other islands of the chain, the people lived in a close relationship with the sea, the men being skilled at hunting and making boats and useful objects from wood and bone. The women produced clothing from the skins of birds, seals, and otters and made the essential waterproof garments, *kamleika,* by intricately sewing the intestines of seals. They were also adept at weaving the coarse elymus grass, found at the water's edge, into baskets and mats with many uses—a skill they have retained.[6]

Atka is now one of only seven inhabited islands in the archipelago. Earlier in 1936 a white trader had been murdered by a native man, an unusual case in that the cause was jealousy rather than alcohol abuse.[7] A cloud of misery still seemed to hang over the village when the *Chelan* dropped anchor in Nazan Bay. When Isobel went ashore with the medical officer only the children showed any signs of welcome. With limited time, she followed a small stream heading straight up onto the hillside overlooking the bay to collect plants.

Ten years later Ethel Oliver, the schoolteacher on Atka, described her pleasure in the vegetation:

> Sometimes we climbed up the flower-clad slopes clear to the bare, rocky, windswept mountaintops. Not totally bare, to be sure, for the tiniest of flowers clung to the rocks everywhere, along with beautiful lichens . . . violets with stems only long enough to put the blossoms above ground, dwarf lupine, dainty short-stemmed rosy primula, a miniature bluebell, and ever so many more. In fact, most of the flowers which we found growing so beautifully large in the lower meadows we also found on the mountains, clinging close to the ground, for the winds really blow across the heights.[8]

In the short time allowed, Isobel raced to collect over fifty species, one of which she had not seen farther east.

Earlier that summer a ship of the Japanese Bureau of Fisheries had arrived in Nazan Bay, foreshadowing tragic events that would overtake the western Aleutians during World War II. The ship purported to need fresh water, but it stayed nearly a week, sending parties of sailors to explore and questioning the storekeeper, Mr. Wheaton, about good harbors on other islands. The Aleut people quietly disappeared from the village while the ship was in the harbor,[9] since the Japanese were generally feared and distrusted throughout the Aleutians. "For years [Wheaton] had noticed how a trapping season always ended with an Aleut trapper's bringing in some tale of having found tin cans with oriental labels on one of the islands, or traps not used in the western world. The whole village would be upset for weeks after hearing those stories."[10] The Atka chief, Bill Dirks, reported the same story about the ship anchoring in Nazan Bay and added that the chief on Attu, Mike Hodikoff, had tried to tell the government that the Japanese were searching for good harbors in the Aleutians but was not believed.[11]

Plants know no boundaries, national or continental. When Isobel was planning her trip, she was aware that Japanese boats sometimes appeared in Aleutian waters. In her wildest dreams she hoped for a chance meeting with a Japanese fishing boat on which she could follow the trail of plants from one continent to the other along the stepping-stones of the Aleutian archipelago. Sailing with the Coast Guard, constantly on guard against poachers in American waters, she had to suppress that forbidden thought.

The administrator on Atka in 1946, Simeon Oliver, had grown up at the Jessie Lee Home where Isobel was billeted during her stay in Unalaska. At the home the boys had played "poachers and revenue cutters" instead of "cops and robbers," the poachers being the fast Japanese boats seeking to outwit the Coast Guard. He also remembered a camping trip when boys from the home had hidden a deserter from a Japanese ship and been terrified as the officers relentlessly searched for him.[12] The Japanese had a sinister reputation throughout the Aleutians that Isobel, with her innate trust of all people, was happy to ignore.

The *Chelan* continued westward from Atka to uninhabited Amchitka to pick up a small party that had been landed there for several weeks' observation of the sea otter population at the island sanctuary. There were traces of poachers suspected to be operating in the area, but they had escaped under cover of fog. Isobel went ashore with a sailor to act as her assistant, and the low, undulating island yielded plants for the presses as well as ripe seeds from the blooms long past in late August.

Captain Keilhorn took advantage of the unusually fine, calm weather to chart a reef noticed the previous summer, just west of Attu Island and close to the border with Russia. For seven days the *Chelan* crisscrossed the uncharted waters beyond Attu, covering an area of twenty thousand square miles. With a comfortable cabin to work in, Isobel also had the freedom of the ship and occasionally watched the operators of the echo sounder at work. The depths recorded varied from more than four thousand fathoms in the Aleutian Trench to a shallow forty-nine fathoms. Having found the high point of the submerged reef, the navigators set about defining its boundaries. As she later reported to the Royal Geographical Society in

London, the unexpected discovery would interest archaeologists as well as geographers, since it showed that the Aleutians once formed a continuous mountain chain linking the two continents, enclosing the Bering Sea in a shallow basin. The submerged reef lying between Attu and the Komandorski Islands is in an area where hydrographic charts had previously marked a deep ocean trench.[13]

While the *Chelan* passed back and forth over the seas, the little island of Attu lay tantalizingly close. The weather, having cooperated beyond all expectation, could no longer contain itself. The wind came up hard from the east, the sky darkened, and the seas began to "show their teeth." Although Chichagof Harbor on Attu offers good shelter, it is small and difficult to find in rough weather. For two days the ship rolled on the heaving Bering Sea, the decks awash in spray, but on the third day patience was rewarded when the skies cleared, even though the sea was still in turmoil. Toward evening it grew calm enough to anchor close to the harbor mouth, and Isobel prepared to go ashore the following morning with a landing party. If the weather should suddenly worsen while they were ashore, the group would be abandoned on Attu until the ship could safely return. For Isobel, with her love of remote islands, the prospect of being marooned on Attu was attractive. Knowing that the stop would be brief, Captain Keilhorn assigned two sailors to help her search for plants. Capturing sixty-nine species, including several emigrants from Asia, unknown on the islands east of Attu, she felt rewarded for her determination and stamina in persisting to reach her goal at the farthest end of the Aleutians.

The people of Attu, away from most corrupting influences, were considered by the men of the Coast Guard to be the happiest and best of all the Aleuts. Even in her limited time ashore, Isobel had the opportunity to meet and photograph several families, including that of Chief Mike Hodikoff. Looking at the photographs of the smiling family groups, posing with obvious pride and pleasure, one feels a terrible poignancy, knowing the fate that is about to overtake those happy people. Six years later, on a Sunday morning in June 1942, while the villagers were attending the Russian church service, Japanese gunboats arrived in the harbor. The people put up no resistance, yet some of them received gunshot wounds, and the American wireless operator, husband of the schoolteacher, was killed. The people were kept under close guard until September, when they were sent to

Japan on a coal carrier. At the end of the war, of the forty-two taken from the island only twenty-four had survived the ordeal. The others, including Chief Mike Hodikoff and his family, died in Japan of tuberculosis, beri-beri, and starvation. Of the several children born there, only one lived.[14]

The Japanese moved quickly to take the neighboring island of Kiska and bombed the village on Atka for three days. The people of Atka were victims of negligent treatment by the American authorities that was almost worse than that received by those who were prisoners in Japan. They were evacuated to an abandoned fish factory in southeastern Alaska, largely forgotten, and so neglected that many did not survive. Their homes and church were burned by the American military to prevent them from falling into enemy hands. At the end of the war, both the Atkans and the people of Attu were brought back to Atka and supplied with material to build new houses. Attu remained under the control of the American military.[15]

With no premonition of what fate had in store for the little island of Attu, the *Chelan* left its anchorage in Chichagof Harbor in a rising gale and began the journey east. The departure for the home port was celebrated on board by a feast, and in spite of the furious waves crashing across the bow and cascading onto the deck and the surging green water pummeling the portholes, Isobel had no difficulty joining in the repast. She had become a true sailor.

Before making a brief return to Atka, there was one last port of call, the sheltered harbor on Kiska, the island reserved as a naval base. Arriving there at the end of August, Isobel filmed the crew executing lifeboat drill, and Captain Keilhorn spent three days calibrating the compass in the harbor. In sunshine, she climbed to the summit of the mountain south of the harbor carrying her flower presses, sketchbook, and paints. Kiska's volcano was enveloped in fleecy cumulus clouds, but Semisopochnoi, the Island of the Seven Mountains, was pale blue in the far distance. "[I] stood there hatless in the wind and sun looking down upon the outspread islands and rocks, and the wide bay with its green walls where the 2000-ton cutter now looked like a child's silver toy." Unable to capture the scene in paint, she watched the eagles and made up verses about "this beautiful forbidden Isle of the Sea, which we were so soon to leave again to her loneliness in the custody of the fox and the fowl."[16]

The cruise on the *Chelan* was coming to an end. Isobel had spent three weeks as the only female member of the crew of ninety sailors. As a memento, they presented her with her personal flag inscribed in black lettering, "Isobel Hutchison, Admiral of the Bering." Using a sewing machine, the quartermaster had embroidered a large gold flower in each corner in place of stars, and where there would normally be crossed swords in the center he had made a design of crossed hairpins. Back in the harbor at Unalaska, the end of the cruise was recorded on both still and cine cameras. Isobel, wearing her best tweed suit and cloche hat, sat in the center with her flag proudly draped over a ship's life belt in front of her. Beside her was Captain Keilhorn flanked by the other six officers, perfectly attired in dress uniforms, white gloves, and swords.

Isobel packed a consignment of plants, her second one from the Aleutian trip, and sent them to the British Museum in London. In a letter to the museum, she described her trip in some detail:

> Though I did not have many hours on the islands, they sometimes gave me a sailor . . . to help me to gather the flowers, and so in this way I was able to bring back from the islands visited (Atka, Kiska, Amchitka and Attu), about 250 specimens. These were mostly the same as the three hundred or so varieties I got in Unalaska, but on Attu . . . I was much interested to get quite a few species which are *different* from all the other islands and are probably Asiatic. . . . It was all intensely interesting and the flowers were very beautiful. I do not wish to be greedy but hope the Museum will be able to make me a grant for the collection this time. I had suggested $500 (one hundred pounds) but if this is too much, though of course it does not nearly cover expenses of travel, I shall be glad to have whatever you think the collection may be worth to the Museum.[17]

Her tour with the Coast Guard was complete, but Isobel was not yet finished with the Bering Sea. At Dutch Harbor on Unalaska, she joined the cruise ship *Mount McKinley*, making its first voyage from Seattle to Nome. Entrusted with money and a commission from the Royal Scottish Museum to purchase Eskimo artifacts, she knew that Ira Rank, owner of the *Trader* in Nome, would be able to satisfy this request. The artifacts provided a valid excuse to visit Nome and renew the deep friendships formed three years before.

It says a great deal about Isobel Hutchison that she would undertake a "cruise" of one thousand miles in each direction on reputedly the world's roughest sea on the spur of the moment to carry out an assignment and to visit friends. Because shipping schedules in the region were so uncertain and her cruise on the *Chelan* was of undetermined length, even had she known of the possibility of sailing on the *Mount McKinley,* she could not have arranged it in advance.

Isobel had a joyous reunion with her friends from the *Trader,* Ira Rank, Pete and Kari Palsson, and big, smiling Gus Masik, who reached for her outstretched hand with "the grip of Alaska." Masik, having left Martin Point, was working a small gold mine tunneled out of the tundra under Anvil Mountain behind Nome. Among fragments of cine film held by the Film Archives of the University of Alaska are a few frames of Gus and Isobel panning for gold together and of Isobel, with a rope tied around her waist for safety, laughing as she descends the ladder at the mouth of the eighty-two-foot shaft to view his latest venture.[18]

Since there was no harbor at Nome, the *Mount McKinley* anchored in the roadstead a mile offshore for three days unloading freight. The lack of landing facilities did not hinder Isobel from spending each day with her friends. Boarding the ship each evening was an adventure in itself.

> Sometimes I was slung aboard in the sort of horse-box in which Nome's long-suffering populace take their first "hop" outside, sometimes I took a perilous leap in the dark from a barge piled with coal-sacks toward a ladder suspended far above me from the ship's prow, its lowest rung missing, its second broken, and its third doubtful. . . .
> But it was worth it all—and more—to see Nome again.[19]

Born out of the gold rush, and now dowdy and dilapidated, Nome was the home port of the "mosquito fleet" that developed when gold fever abated. Small trading vessels crisscrossed the icy waters of the Bering Sea, in and out of Siberian ports, in the early part of the twentieth century until the Russian Revolution made trade uncomfortable and dangerous. That coast, from Kamchatka to East Cape, was a source of wealth in furs and ivory that caught the imagination of men for whom adventure was as important as commerce.[20] In 1933 Isobel had spent four months in the sole company of men who were part of the mosquito fleet, and their love of the northern frontier infected her. The return to Nome renewed the sense of glorious freedom she had experienced on her earlier journey.

In Nome Isobel also met Mrs. Greist, the missionary from Barrow, waiting to go "outside" from Alaska.

Right royally did our friends entertain the pair of us, doing the honours of
the new Nome now uprising like a phoenix from the ashes of the great fire
which destroyed most of the famous old mining "city" in 1934. We climbed
mountains, visited gold-dredges, motored far out across the blue and gold
moorland, past foaming rivers, till the track ended. Then, having descended
into the heart of the earth to visit Gus's mine, we had coffee (prepared by
himself) in his shack, washing up the dishes for him as of old. The party
of six were thereafter entertained to dinner by himself and Captain Pete
at "Cavey's" Restaurant in the town, where amongst the Alaskan dainties
dispatched were chicken, ptarmigan, pie, soup, tea, beer and coffee![21]

On the thirteen-day voyage to Seattle, Mrs. Greist, whose contact with
leading anthropologists had given her expert knowledge during long service
in the North, helped Isobel label the artifacts they had chosen together for
the Royal Scottish Museum. Off Unalaska, a hurricane caused havoc among
the small boats and sent their ship scurrying for shelter in Dutch Harbor.
While anchored there, Isobel climbed to the summit of the mountain on
the island in the harbor with a fellow passenger. "For the first time I experienced the curiously exhilarating sensation of literally wrestling with the
wind, and the wind won easily, lifting me right off my feet several times,
so that I was obliged to hold on to the steep mountain-side by hands and
knees. On our homeward way we were caught and drenched to the skin in a
torrential downpour."[22]

What had begun five months earlier as routine travel by scheduled
carriers took on the flavor of adventure by virtue of Isobel's willingness
to grasp any opportunity that presented itself. Flexibility was only one
of the characteristics that made her a successful traveler. Interest in her
surroundings and enthusiasm for each new experience endeared her to the
people she met along the way, forever widening her circle of friends and
correspondents. The young woman who twenty years earlier had confided
to her diary the hope of making one new friend now had friends scattered in
out-of-the-way places across the North. With self-confidence growing out of
her travel experiences, she was equally comfortable spending a week on an
island of strangers or three weeks as the only woman on a naval vessel with
nearly a hundred sailors.

Being "employed"—with remuneration still in question—as a plant
collector for the British Museum gave Isobel Hutchison a serious scientific
purpose. Since the eighteenth century, botanists had made collections in the

Aleutian islands and the Pribilofs, and she was familiar with the names of many of them. She carried with her the most recent and complete catalog of Aleutian flora by the Swedish botanist Eric Hultén, based on his thorough investigation in 1932, when he visited fourteen of the islands. The second edition of his book, published in 1960, lists her among the plant collectors and in the bibliography. Because the individual entries for each plant are based on Hultén's own collection and on the collections in a dozen different herbaria, which did not include the British Museum, there is no reference to the specimens she collected. She is clearly the woman, of five listed as collectors, who made the greatest contribution, the other four having collected only single specimens from Unalaska.[23]

Isobel Hutchison's book *Stepping Stones from Alaska to Asia,* about her Aleutian journey, was published in 1937 and reissued in 1942 in a "cheap" edition titled *The Aleutian Islands,* because of increased interest in the area after the Japanese invasion. In the preface to the second edition, which eliminated the poems and the appendix listing the botanical specimens collected, she refers to it as a "little record of a happy summer." In alluding to the tragic events that occurred on the islands where she had known joy and peace, she wrote with the serene optimism of one who has survived her own battles and retained an unshakable faith in the Creator. "But wars, after all, are transitory; it is the flowers that last."

As a conscientious plant collector, Isobel leaned heavily on the use of Latin botanical names to describe the flower-decked hillsides of the Aleutians. Although she was well informed on the history of the area and enjoyed the Russian flavor, she tended to balance the serious side with whimsy. The review in the *Times Literary Supplement* mildly chided her for this: "Miss Hutchison invites the comment that travellers' tales should not be too sweet. . . . she too often obscures the misty beauties of the Aleutians in a Scotch mist of the more whimsical variety."[24]

It is an indication that Isobel was allowing herself to take her scientific work seriously that she published a brief article in *Nature,* one of Britain's foremost scientific journals. The article, "Plant Collecting on the Pribilof and Aleutian Islands, 1936," described the extent of her travel and noted that she had collected some two thousand specimens for the Natural History section of the British Museum.[25] A review of her book on the Aleutians appeared later in the same journal. After praising the author for the care with which she followed the dictates of specialists who worked in the region, in both botany and ethnology, the writer of the review signified that he himself was at least as experienced in northern travel and perhaps knew the subject more intimately than she did. Emphasizing Isobel Hutchison's amateur status, he summarized her quietly remarkable accomplishment:

> The author is an energetic Scottish woman, apparently of middle-age—an ardent and experienced, if technically untrained, botanical collector—with grit and perseverance as outstanding attributes. Now here lies a poignant indication; she has recently visited several of the least accessible regions in the world alone and without "expedition press" or acclamation as a "heroic explorer," which suggests once more that the epoch of such things and people is drawing to its natural close, and that its place is being taken by one of serious scientific investigation and honest travel.[26]

Within the review quoted above is the key to understanding the style of Isobel Hutchison's travel writing. She had gained entrance to the closed world of Greenland on the strength of her status as a plant collector working for members of the Royal Horticultural Society. Her book *On Greenland's Closed Shore* contained less about plants than about the whole experience of living in an isolated native community, rarely visited by outsiders. She had used that experience to persuade Kew to authorize her to collect for them on the trip around Alaska. Having volunteered to collect artifacts for Cambridge University as well, she found that the backing of two venerable institutions was valuable when her travels, as reported in *North to the Rime-Ringed Sun,* turned into a much greater adventure than she expected. Before setting out for the Aleutians, Isobel had secured the endorsement of the British Museum, whose reputation opened avenues of travel that would have remained closed without it. In this third foray as a plant collector, her knowledge of the botanical literature and her experience with plants made her highly skilled at her work. She took collecting seriously, but to write an account of her travels to be read by the general public required a light touch. *Stepping Stones from Alaska to Asia,* perhaps the least successful of the three travel books, tried to straddle both worlds, the scientific and the popular.

Isobel Hutchison would never have characterized her journey to the Aleutians as wildly adventurous. It was, however, an exploration of a little known, seldom traveled region, and to accomplish as much as she did required patience, a sense of humor, and the courage to accept any transportation, accommodations, and food available. Her journey was not over. On Attu she had seen plants native to Asia. With the persistence of a dedicated amateur, she must continue to follow the clues she had seen.

13

Around the World and Home Again

Isobel had two weeks in Seattle before meeting her sister in Vancouver in October 1936. Both Coast Guard officers, Commodore Dempwolf and Captain Kielhorn, and their wives entertained her with drives around the area. She gave a party for the officers of the *Chelan*, showing the color film she had shot, and there was a farewell dinner aboard the ship. The sister of Pete and Kari Palsson, her shipmates on the *Trader*, invited her to visit Wenatchee, and there she also caught up with Alan May, whom she had met in Unalaska. He arranged for her to lecture at the Archaeology Institute in Wenatchee. Visits to university and private herbaria and to friends from the Yukon and Alaska easily took up the time. Her diary is filled with examples of kindness received and her own efforts to return hospitality. Clearly, Isobel was someone people enjoyed.

Late in October Isobel and Hilda settled into a tourist-class cabin aboard the *President McKinley* and sailed from Victoria on the two-week crossing to Yokohama, Japan. Isobel had the novel *Gone with the Wind* to occupy her, but she was on deck to watch as they passed close to the Aleutian chain at Unalaska. After the ship was beyond distant Attu, she felt the effects of the Bering Sea, rough as usual. The visit to the Aleutians had been a working expedition, albeit with plenty of free time between intense sessions of plant collecting on the islands where she landed. The journey home, across half the world, was a holiday with Hilda, and no published articles resulted from the trip.

They spent three weeks in Japan, staying mainly in villages in the hills, which Isobel thought the loveliest of all the lands she had seen. Crossing through China, on December 10 they boarded the Trans-Siberian Railway at Vladivostok for the weeklong ride to Moscow, for which Isobel used a word rarely found in her vocabulary: "boring." Not allowed to leave the station in Moscow, they continued directly through Poland to Berlin, where they spent Christmas. Passing through The Hague and Amsterdam, they took the ferry from Harwich to London, arriving on the last day of 1936. The activities for that day show two facets of Isobel's character, the practical and the spiritual. She shopped for new clothes at the post-Christmas sales and attended the watch night service, beginning just before midnight on New Year's Eve, at Saint Paul's Cathedral.[1]

Soon after arriving home Isobel published three long feature articles in the *Times* about the Aleutian and Pribilof Islands and worked on her book on the Aleutians, to be published in the autumn.[2] She entered into negotiations with the Royal Geographical Society regarding both a lecture and a showing of her color film in the spring. Although she had previously informed them of her Greenland lecture, they had not offered her a place on their program. Through Mr. Ramsbottom of the British Museum she heard that there was now some interest in having her speak.

Negotiations with the Royal Geographical Society were not as simple as with other organizations where she lectured. When she told them that she would be speaking in Glasgow, Dundee, and Edinburgh in March, the RGS made it clear that the lecture in London must be on original material not used in Scotland, since they would wish to publish a paper resulting from it. This raised concerns with her publisher. If the paper was long and appeared before her book, it would hurt sales. At last it was settled that Isobel would show her film on April 26 at a film night and lecture on May 3 on a topic not covered in Scotland and show slides not used there. She could add a short piece of film to the lecture, since the two audiences would be different. The paper for the *Geographical Journal* of the RGS, on the discovery of the new reef off Attu Island, was only three pages instead of the suggested ten or twelve, but it was augmented by photographs and by charts given to her by the Coast Guard.[3] It appeared in December 1937, following the publication of her book in October, and thus Isobel satisfied the requirements of both the RGS and Mr. Walter Blackie, her publisher.[4]

Just forty years earlier, Isabella Bird Bishop was the first woman ever to address the exclusively masculine Royal Geographical Society, lecturing on her travels in western China.[5] In 1892 the admission of fifteen "well-qualified" ladies, including Mrs. Bishop, to membership had caused a furor that boiled over in the columns of the *Times*. It was led by Lord Curzon, fresh from travels in Africa, and Admiral Sir Leopold M'Clintock, the noted Arctic explorer. Lord Curzon was vociferous in his attack: "We contest the general capability of women to contribute to scientific geographical knowledge. Their sex and training render them equally unfitted for exploration; and the genus of professional female globe-trotters with which America has lately familiarized us is one of the horrors of the latter end of the nineteenth century." The ruling to admit ladies was voted down, although those already elected were permitted to remain.[6]

The Royal Scottish Geographical Society had been open to women from its inception and prided itself on drawing its membership from "all grades of society—noblemen, country gentlemen, men of science and letters, professionals and schoolmasters, merchants and shippers, clergy, lawyers and physicians, women as well as men."[7] The Royal Geographical Society in London, having finally overcome the fear that its scientific status would be threatened if women were admitted, accepted them as full members in 1913.[8] Isobel was made a fellow of the Society in 1936 and presented it with the United States Coast Guard charts of the Aleutian Islands as well as a large number of photographs and a watercolor.[9]

The films Isobel showed at the film night are preserved in the Hutchison collection of the Alaska Film Archives at the University of Alaska at Fairbanks. In muted colors, the photographs of flowers are disappointing, action provided only by the ever-present wind blowing through fields of bloom set in magnificent mountain scenery.[10] The film of fur seals in the Pribilofs contains plenty of action, since the bulls are constantly on the move controlling their harems while their many wives lollop endlessly after their pups. Isobel crept close to the herd, causing them to move en masse toward the sea, some diving in to swim away, others returning up the beach. She had been shooting cine film since her year in Greenland, and considering the conditions and the equipment, the results are commendable.

A letter from Vilhjalmur Stefansson is a reminder that the Aleutians were a relatively unknown corner of the world. "What I am most interested in is information about the Aleutian Islands, there is less of value printed on them than on practically any other segment of the inhabited north."

Late in November Stefansson also wrote to thank Isobel for his copy of *Stepping Stones from Alaska to Asia,* and his letter had news of Gus Masik. "Masik just went through eastbound planning to visit you in Scotland. I had only one glimpse of him for we did not get in touch until the afternoon of his last full day in New York. I am not exactly hoping to see him on his way back this way for he wants to return to Alaska by way of the Trans-Siberian Railway. It is now apparently difficult for those who speak Russian to get a visa—better to keep out one hundred genuine tourists than admit one spy."[11]

The year 1937 was filled with happy events. Isobel had paintings accepted by the Royal Scottish Academy as well as shown at an exhibition with other artists in London. The Thalbitzers from Denmark were in Edinburgh and were entertained at Carlowrie along with the Danish consul. All three Hutchison sisters—Isobel, Hilda, and Nita—were in London for the coronation of King George VI. In November Isobel had a full slate of speaking engagements, including Cambridge, Studley College, and the Forum, a women's club in London, before she returned to Edinburgh just in time to attend a lecture at Usher Hall by Grey Owl, an Englishman whose adopted guise as a Canadian Ojibwa Indian had been accepted everywhere without question.

Just over fifteen years earlier Isobel had been living as a student in London, staying at a women's hostel where the discovery that her Australian college friend was a prostitute caused a mental breakdown in the naive and confused young woman. After more than ten years of independent, "hard" travel in northern regions, she exuded confidence and energy. Using the Royal Geographical Society on Kensington Gore as her address, she was only a mile from the hostel where she had lived in 1920, but she had circled the globe to arrive there.

At the beginning of December Gus Masik arrived in Scotland. He stayed in an Edinburgh hotel, but Isobel took charge of every day, showing him the sights, taking him to scheduled events, and introducing him to all her friends and relatives. He went along with her when she lectured at the Philosophical Society in Glasgow and met her publisher there. She even took Gus, the nonbeliever, to a service at Saint Giles's Cathedral. From photographs taken during his visit, it appears to have been a happy time for all concerned. After two weeks Isobel saw Gus off on the train to Hull, to sail on to Estonia for a reunion with his sister.[12] She wrote on the last day of December, "A lucky and happy year." It had been eventful from start to finish.

In the summer of 1938, after attending an ethnographic conference in Copenhagen as a delegate from the Royal Scottish Geographical Society, Isobel visited Estonia. Connections with the Baltic seaports and Russia through her great uncles on her mother's side gave her an interest in that region, with its fluid borders and distinctive peoples, but her friendship with Gus Masik was the spark. Gus's sister Anna was expecting her, and when Isobel alighted from the train from Tartu "a smallish thin man [Anna's husband] in best tweed suit and large brimmed felt hat coming down over ears" stepped forward to meet her. As she tried out an Estonian phrase learned for the occasion, the German-speaking station master offered help. Isobel voiced her intention of returning to Tartu on the evening train, but it was explained that she was expected to stay for two weeks!

A cart pulled by a glossy horse, the seat a bundle of straw covered with an embroidered cloth, jolted them over the bumpy country roads to the farm. "People working in the fields stop to stare at the strange looking lady with a *hat,* green waterproof dust coat, cocked arm in arm with the driver." His arm sometimes tightly encircled her waist to keep her from falling off the wagon. At the farmhouse Anna, a small thin woman whose only resemblance to Gus was her steady blue eyes, stood with family and neighbors crowded around the door, all frustrated by the lack of a common language. Without language, Isobel resorted to showing photographs. "A picture of Gus in Scottish dress prompts the remark, 'August my bridegroom.' I show them photo of *Punch* my dog and say *he* is my bridegroom, and much laughter."

With the arrival of three young girls who spoke German, the questions began again. When Anna asked why Isobel did not marry August, she explained that he was like a brother to her. Anna had evidently questioned Gus on the subject and had been told that Isobel was too "high" for him. Isobel, growing steadily hotter from the questions and the heat of the kitchen, was ushered into the outer room where the table was laden with food. In spite of her accepting primitive conditions and strange food without comment in Greenland and Alaska, the peasant farmhouse kitchen, with its dirt floor, the slabs of food, a plate of raw eggs, the lack of cleanliness, the heat, and the crowds of people almost defeated Isobel. The pleasure she felt at arriving in what seemed like picture-perfect Estonian country surroundings had evaporated. In spite of Anna's best efforts it was undoubtedly a relief to both parties when she returned to Tartu after two nights.[13]

Isobel had been in places more primitive, and probably dirtier. She had struggled with language problems on other occasions. Without knowing the language she felt like a prisoner in the little farmhouse, her sense of freedom stifled. Her emotional connection with Gus, coupled with the presumption of familiarity from Anna, who referred to her as "sister," had made her uncomfortable and stiff. Although she had many friends on different levels, few penetrated the barrier of reserve that protected her strong sense of privacy.

Her manuscript of a book on Estonia was never published, but the *Times* printed an article that showed her strength as a journalist and her boldness as a traveler. Beginning with historical background and a brief description of the beauties of the capital, Tallinn, Isobel went on to describe her exploration of Estonia's border with Russia. On both sides, armed guards with binoculars watched each other across a strip of no-man's-land. With an Estonian woman journalist she had met in Tallinn, she followed the line of the frontier, which divides the waters of a large lake. Both Russians and Estonians had been killed in recent instances of illegal border crossing in the area. Isobel and her friend took a starlight swim in the lake and later learned of yet another desperate frontier drama that had taken place nearby the previous night.[14]

This unrest was a symptom of the international tension developing throughout Europe at the end of the 1930s. The uneasy year of peace that followed the signing of the Munich Pact with Hitler was finally shattered on Sunday morning, September 3, 1939, when Britain declared war on Germany. Isobel was walking the mile to church in Kirkliston an hour after the fateful announcement, on a day of thunder with bright intervals, when the siren at the village distillery blared an air-raid warning.[15] It was a false alarm, but it marked the beginning of six years of disruption to lives throughout Britain and the end of Isobel's travels in the North.

For everyone in Britain the war effort was the priority. Having registered for war service, Isobel and Hilda were called up almost immediately and sent to Liverpool to work as censors. Isobel dealt with letters in Danish and Norwegian, while Hilda worked in Italian. It was the first real "job" for either sister, working regular hours and sometimes weekends and living in "digs" with a landlady. After six months they were transferred to Glasgow, living outside the city in the village of Killearn. Nita, now a widow, had taken refuge with the housekeeper from Carlowrie at Bettyhill on the north coast of Scotland. Carlowrie became a billet for officers from the Royal Air Force, having been rejected as too close to an ammunition dump to receive evacuated children. The house was maintained by a skeleton staff, and the sisters visited it as often as possible. Isobel continued to give lectures around Scotland for the Ministry of Information and was also appointed a justice of the peace at the beginning of the war.[16]

Although the German air raids concentrated their fiercest assault on the homes of the shipyard workers on Clydeside, wrecking forty thousand homes in two nights of bombing in March 1941, the area around Carlowrie received only one heavy shower of incendiary bombs, which fell harmlessly on the "bings" or slag heaps of the oil-shale mines of West Lothian.

After two years the censorship work ended, the RAF personnel suddenly vacated Carlowrie, and Isobel returned home. Taxes were now taking half her income and money was scarce. She looked for more war work with the Ministry of Information, but nothing came of it. After the Japanese invasion of Attu Island in the Aleutians in June 1942, with her unique knowledge of the area the BBC asked her to give a radio talk as part of the Home News, ritual evening listening for most Britons. She had made her first broadcast of poetry in 1928 and done occasional talks on her travels, but this was the beginning of regular broadcasts, both giving talks on current affairs for the general public and appearing on the Children's Hour.[17]

An article "The Riddle of the Aleutians," submitted to *National Geographic* soon after her return, had also become a timely subject and was published in December 1942, with photographs supplied by the United States Navy giving it an air of immediacy. Isobel had first published in *National Geographic* in 1928 with an article on her walk across Iceland, but she had not been successful in recapturing the magazine's interest until the accident of history gave her a new entrée. The editor, pleased with her story on the Aleutians, suggested that she write an article on Scotland in wartime, which was published in June 1943 and followed a year later by "Wales in Wartime."[18] Over the next dozen years the American magazine, with its familiar yellow cover and pages of glossy photographs, published eight more of her articles, making a total of twelve.

The two wartime articles have an optimistic tone, with an emphasis on the lighter side of the war. While the Scottish article mentions the first civilian casualties, occurring in Orkney, and the savage bombing of Clydeside, Isobel used her skill in recording dialect to show the other side of wartime life, using sometimes comic stories of children evacuated from the Glasgow slums to the relative safety of rural Scotland.

A summer tour of Wales in 1943 provided the material for "Wales in Wartime," published by *National Geographic* the following year. Pastoral and mountainous, Wales was fulfilling its old role as "England's refuge room," with evacuees and refugees filling every available space, especially in the rural areas where food rationing was more relaxed. After traveling around Wales on crowded buses and unreliable trains, Isobel turned the trip into a holiday by hiking alone over the hills, knapsack on her back, and visiting her lifelong friend Medina Lewis. Medina and her mother contributed to the war effort by collecting medicinal herbs and drying them in the tannery that had housed the family leather business.

With her first foray into Greenland, Isobel had escaped the silken ties of duty by which widowed Victorian mothers held their unmarried daughters in bondage. While Isobel went to the outermost corners of the earth, Medina was tied to her home in Newtown as companion to her mother, traveling vicariously through her friend's letters and waiting for her own turn to come.

During the war, Isobel continued to receive letters and parcels of chocolate from her friends in America. Under the terms of the Lend Lease Act in 1941, the United States Coast Guard ship *Chelan* was transferred to the Royal Navy. Captain Kielhorn wrote from Washington: "A couple of weeks ago I went aboard the *Chelan* for a last cruise, and with my own hands I compensated her compasses and I saw too that she was in good shape— fit for duty different from any we could have dreamed five years ago. . . . I am sure she will give a good account of herself, for she was always a lucky ship." He was interested to hear from Isobel that his ship had rammed and sunk a large Italian submarine and that its crew had rescued thirty-six survivors.[19]

The end of the war brought a gradual return to normality, though not to prewar standards of living. Wartime inflation played havoc with incomes based on investments made before the First World War. The intention of the original trust set up by Isobel's father was that the expenses of Carlowrie should be covered by the rent from the tenant farm. Isobel had a good relationship with Peter McGowan, her tenant farmer, and was loath to increase the rent to keep pace with inflation. The staff at Carlowrie were like part of the family, but age was taking its toll. Greater mobility of the population, emigration to Canada and Australia, and a new independence among working people in postwar Britain made hiring replacements more difficult, and thus the staff was reduced over the years.

There had been few changes in Carlowrie since the death of Mrs. Hutchison, and during Isobel's long absences combined with wartime restrictions it had suffered twenty years of neglect. Isobel wrote to Reg Orcutt, "Our house, which is to get electric light installed we hope in 1951, will be a hundred years old next year, so we hope to celebrate the centenary and still be in it, though I'm not so sure, it is so hard to keep going nowadays with the rises in price and wages. However we are now building up instead of pulling down and all the dry rot damage I hope is repaired and some improvements added."[20]

The large walled garden that today encloses a perennial border surrounding a perfect lawn was previously a kitchen garden, and with economic necessity it became a market garden. Isobel scrambled from writing articles and preparing broadcasts to picking flowers and fruit for the Edinburgh market. Her time was divided between attending council meetings of the Royal Scottish Geographical Society and helping the gardener saw dead trees into firewood or spread loads of gravel on the driveway.[21]

Fortunately there was still time to travel. *National Geographic* requested another article on Scotland just after the war ended. Restrictions on travel were still so tight that Isobel had to work through the Ministry of Information to get a car for the magazine's photographer to accompany her. Her next writing assignment was in Denmark, where she had been invited to lecture by the British-Danish Society after being awarded the Danish Freedom Medal in 1946.

The weather during January and February 1947 was exceptionally severe in Britain. Snow and ice covered the ground continuously, and some areas saw no sunshine for weeks at a time. Adding to the misery was a coal shortage. During that year the country was hit by an economic crisis when an outflow of currency threatened to bankrupt the nation. Stringent measures were adopted to stanch the wounds. Dollar imports, including food, were cut to a minimum, and food rations were reduced below the wartime level.[22] The British pound was devalued, and the amount of foreign currency a private citizen was allowed to buy would barely permit travel abroad.[23] Two things made it possible for Isobel to visit Denmark. In addition to her invitation to lecture there, the National Geographic Society, accepting her proposal for an article on Denmark, provided her with a photographer, Maynard Williams, and a car. Late in April she and Hilda sailed from Leith to Copenhagen, moving from the bitterness of the British winter to the sunshine of a Danish spring.

Although Denmark had suffered from occupation by the enemy throughout the war, the country was already the leading exporter of food in Europe. The Hutchison sisters relished the protein-rich meals that British people had hungered for since 1939, as well as enjoying their contacts with Danes, many of whom spoke English. Among those Isobel met was Baroness Karen Blixen-Finecke, the traveler who wrote under the name Isak Dinesen, in her castle at Hesselager. She renewed friendships from her Greenland days, being entertained by the Thalbitzers, the Mikkelsens, Dr. Bentzer, and others as well as giving a party for them.[24] In spite of the agreeable conditions, the article "Two Thousand Miles through Europe's Oldest Kingdom" is darker in tone than Isobel's wartime articles on Scotland and Wales. The horrors of the occupation and the lingering memory of deaths of individual members of the resistance movement cast a shadow over the flowering springtime.

More honors were in store for Isobel. On June 30, 1949, the University of St. Andrews conferred on her the degree of doctor of laws for her work as northern explorer, botanist, writer, lecturer, and artist. Modest and unassuming, Isobel confessed to a reporter who interviewed her for the Edinburgh newspaper, "I really can't think why they are giving me this degree."[25] It was the occasion to buy a new hat, black trimmed with red roses, even though just a few months earlier Isobel and Hilda had needed to sell cameras, jewelry, and some of their china to raise cash. Medina came from Wales and accompanied Hilda to St. Andrews for the great day. Isobel was asked to reply to the toast to the graduates. "I certainly do tremble at this!" But she rose to the occasion with a witty speech. She was the only woman to be honored; the men included American ambassador Lewis W. Douglas and Sir John Gielgud, the actor. Her three days at St. Andrews were a high point in her life. Congratulations on this honor poured in from friends across the world.[26]

It is tempting to label Isobel Hutchison a scientist, but it is a label she herself would have rejected. She was well aware of the limitations of her botanical training, despite being an orderly and meticulous collector. As such she was one of a select group of amateur plant lovers who have added—and continue to add—to the knowledge and understanding of the world's plant ecology.

Formality being the convention in Britain, Isobel was henceforth referred to as Dr. Hutchison. This has led to a misunderstanding of the true extent of her achievements. In *Wayward Women,* a compendium of outstanding women travelers, her travels are well summarized, but because of the title in front of her name it was assumed that the journeys were made by a professional doing her job.[27] More remarkable is that she accomplished them as a gifted amateur and independent scholar, with limited formal training but with a deep love of learning in many areas and with the persistence to reach any destination she chose.

In addition to considering herself an amateur, Isobel Hutchison was consistently modest about the extent and nature of her travels. The gift of a book from Stefansson, who was then compiling another book on Arctic exploration, prompted this response: "Our [RSGS] editor Dr. Kenneth is at present busy . . . writing up something about Scottish arctic explorers for your book on the Arctic. He has told me he is putting in a short bit of biography about myself! (But I told him he had better not as I am only a traveller and no explorer!) I am afraid I have no real claim to be included amongst the great ones there."[28]

Although there were no more major travels resulting in books, writing continued to be Isobel's source of income, and her articles appeared in many publications: *Blackwood's Magazine,* the *Scotsman, Cornhill Magazine,* the *Scottish Field,* and the *Polar Record.* As a beginner in 1910, writing mainly poetry, she found that rejection slips far outnumbered published material. By the 1950s nature and travel articles had become her forte, and it was rare that anything she wrote was not accepted.[29] An article for the Archiv für Polarforschung in Kiel, Germany, resulted in Isobel's receiving a fellowship diploma in 1951.[30]

Among the articles Isobel wrote for *National Geographic* in the 1950s were three based on her "strolls," a form of travel she enjoyed all her life. The first solo walk from Carlowrie to London followed a scenic route, rather than the most direct, and stretched over five hundred miles. Starting out on March 1, 1948, she had thirty-eight days to cover the distance before she was due to keep an appointment in London with the photographer from the magazine. Together they retraced her route by car, photographing points of interest. Hilda walked with Isobel for the first eighteen miles, but thereafter she was on her own. In tweed suit and brimmed felt hat, with a small knapsack on her back, when possible she chose the public footpaths over the hills away from the main roads, enjoying in her mind the company of all the writers who had preceded her along those byways, from Boswell and Dr. Johnson in the north to Shakespeare in the south. Part of the inspiration for the walk was supplied by the heroine of Sir Walter Scott's novel *The Heart of Midlothian,* who walked to London barefoot to plead for her sister's life.

Although Isobel was intent on walking most of the way, she was not a fanatic. When rains descended in torrents she would gladly accept a lift, or when the road led through a heavily built-up area she would take a bus to the outskirts of the town. Always adept at dialects, her stories in the local accents added to the good humor of the article.

The following year *National Geographic* published "A Stroll to Venice," about a walk Isobel began in Innsbruck, Austria, crossing the historic Brenner Pass and the higher passes of the Dolomite Mountains. Carrying a heavy pack, she made that walk in 1950 at age sixty-one. She chose to begin in the middle of May to allow time for the snows to melt in the high passes, although with the late start she had to face the uncomfortable summer heat of Italy. To avoid the sun, where possible she kept to footpaths through the woods. Her thoughts were peopled with historic and literary figures, including Browning at Asolo. She was aware also that Asolo was the home of the noted contemporary traveler Freya Stark, who owned a school for silk weaving.

Stark, one of the few women to receive the gold medal awarded annually to an outstanding traveler by the Royal Geographical Society, was riding by mule through seldom-visited corners of Afghanistan and Persia about the same time Isobel was in Greenland. Stark was not at home when a friend, Madame Cavaliari, took Isobel to photograph Stark's garden, so they did not meet, although each was aware of the other's writings.[31]

Isobel made the third of the solitary walks in home territory in 1952, going from Edinburgh to John o'Groat's, the northernmost tip of Scotland. In planning the route she included ferries whenever possible to add variety to the walk. Some of the ferries she had planned to take no longer existed, just as the occasional hotel marked on her map had ceased to operate. When this happened she was always ready to improvise.[32]

When she made this walk Isobel was in her sixties, and she appears in many of the photographs that accompany the article, small and trim, usually wearing the same tweed suit, beret, heavy stockings, and stout walking shoes. Her interests were mainly historical and literary, yet she took the opportunity to descend a three-hundred-foot mine shaft to the coal face at Brora, and she also commented on the new experimental power developments in the north of Scotland, one using peat and the other nuclear energy.

Isobel's last article for *National Geographic,* published in October 1957, was "Poets' Voices Linger in Scottish Shrines." On this journey she was able to share her love of Scotland and Scottish poetry with Kathleen Revis, a young photographer from the magazine. For Isobel, familiar with both the land and the literature, houses and hamlets in every corner of the countryside were alive with the writers who had lived there, and the article is peppered with quotations and anecdotes.

Of the twelve articles in *National Geographic,* half were on Scotland, the subject she knew and loved best. Having written so much about the relatively small country over the span of fifteen years, there was little left to say, and her suggestions for articles about a pony trek in Galway or a visit to Corsica did not catch the attention of the new editorial staff.

When Isobel gave a talk to the Lyceum Club in Edinburgh in 1955 titled "Some Women Who Did Not Stay at Home—Women's Place in Exploration," the audience may have expected to hear something of her own travels.[33] Instead, they were introduced to women whose travels ranged across the world. Isobel was well aware that she was only one of a select company of fearless women who had followed their dreams to extraordinary lengths throughout the centuries. Her own travels reflected the choices of some of the women she presented.

From her experience of the north coast of Alaska, she told them the story of Anarulunguaq, the young widow from Greenland who traveled with Knud Rasmussen across Arctic North America, providing the essential service of sewing fur clothing for the expedition. She was the first Inuit woman to travel widely and to visit all the tribes of her kinsmen. Isobel began her talk with a fitting quotation from an Inuit *angakok,* or shaman, whom Rasmussen had met on the journey: "If a woman is strong in mind, she cannot be outwitted by a man; and if a woman is nimble and fleet of foot, she cannot be outstripped by a man. Therefore do not look down upon woman. She who seems the weaker is often the conqueror."

After describing travelers from earlier centuries, Isobel turned to examples from the twentieth century using Marjorie Tiltman's *Women in Modern Adventure,* which included a chapter about her own travels titled "Searching for Flowers at the 'Top of the Map.' "[34] Published in 1935, this book was one of a series of works on modern adventurers, many with the words "heroes" and "adventure" in the title. Although most contained brief sketches of the famous, such as Lawrence of Arabia, Stefansson, and Amelia Earhart, there were also books devoted to scientists, to heroic acts in everyday life, and even to pioneer motorists. The book on women outlined the adventures of eighteen women ranging from solitary wanders in Tibet to deep-sea diving and from prospecting to auto racing.

The deep-seated human need for people to admire made the series popular with the reading public. The quintessential attraction of travel literature is that readers can experience vicariously places and situations not met in everyday life. Travelers' tales run the gamut from heroic survival against impossible odds to personal interaction with exotic cultures. The success of the stories depends on the writer's skill in drawing the reader into a foreign world. Isobel Hutchison, with the eye of an artist, takes us into the northern landscape and, particularly in Greenland, introduces us to the native people. With her unchecked facility with words, she was perhaps more successful at telling her stories in person than on paper.[35]

From the first of her travels, Isobel always returned home with photographic slides and gave entertaining talks to groups in the local church hall. Once she had overcome her initial shyness, public speaking came easily without a script or notes.[36] Demand for her as a lecturer grew to include learned societies, women's clubs, and scientific institutions. All together she gave more than five hundred lectures, paid and unpaid.[37]

The fees Isobel received for lectures were sometimes used for charity in an imaginative way. For example, on her summer visit to Greenland in 1935 she went to Umanak for the dedication of a new stone church and presented a bell purchased with the fees for lectures on Greenland given in Scottish towns and villages.[38] In her charities, when possible Isobel liked to involve others to spread the impact. On her first visit to Palestine, by chance she had collected cyprus and oleander seeds that germinated and grew into seedlings in the greenhouse at Carlowrie. The sale of these little plants helped support the hospice of St. Andrew's Church in Jerusalem and continued for forty years.[39] A cry from the heart of explorer Ejnar Mikkelsen in eastern Greenland after influenza killed a large part of the adult population brought an immediate response from Isobel, with enough money collected from friends, and possibly from lectures on Greenland, to support two Greenlandic children. Mikkelsen at that time had 135 orphaned children in his care.[40]

Isobel and Hilda continued to take short holiday trips together to Amsterdam, the Riviera, and Italy. Isobel was always attracted to small, uncrowded islands. Visits to Corsica and Elba in the Mediterranean and to the Frisian Islands and Heligoland in the North Sea, off the coast of Germany, became the subjects of short articles. In the same vein, she led an island cruise to Shetland for the National Trust for Scotland, calling at two of the tiniest inhabited specks of rock in the waters around Scotland, Fair Isle and the island of St. Kilda's.[41] Just after her sixty-fifth birthday she took a trip more in keeping with her solitary nature when she bicycled from Carlowrie to Bettyhill on the north coast of Scotland. By taking the train for one short section because of continuous rain, she made the trip in a week, cycling up to forty miles a day, and returned using a combination of bicycle, boat, and train. The appeal of such a journey for Isobel, despite hilly terrain or threat of rain, would be the glorious feeling of freedom.[42]

By the 1960s the days of Isobel's travel had passed, but the need to earn money was still pressing. Produce from the estate—including bunches of snowdrops in the early spring, plums, strawberries, and peaches in summer, and grapes from the greenhouse—helped pay for the upkeep on Carlowrie. She also shared the bounty with friends. The son of the family solicitor remembers: "I can recall as a small boy that two or three times a year on Saturday mornings an old car would arrive at our front door and this tiny, frail figure in the back would direct the driver to hand in to my parents a basket of fresh vegetables and fruits from the kitchen garden or in the winter it might be a pheasant."[43] With the same noblesse oblige, the Hutchisons marked the retirement of the couple who had served for many years as gardener and housekeeper with the gift of a silver tea service. By this time gaps in the housekeeping staff often left Isobel in charge of cooking and laundry.[44]

In the years following her visit to the Aleutians, Isobel had become a public figure as a broadcaster on the BBC, a regular contributor to magazines and journals, and a frequent lecturer throughout England and Scotland. She had received both an honorary degree and medals for her work, undertaken as a true amateur—for the love and pleasure of it. But for the intervention of World War II, she would almost certainly have made further journeys to distant places. The east coast of Siberia, from Kamchatka to East Cape, was high on her list of destinations. With friends in the United States Coast Guard and a favorable political situation, it is tempting to speculate on the outcome. She was at the zenith of her abilities when war was declared, and the six years lost were only part of the disruption. Without new and intriguing adventures to write about, her name gradually faded from public view.

14

The Blessings of Friendship, the Curse of Old Age

"All the best days of life slip away from us poor mortals first; dreary old age and pain sneak up." The words from Virgil's *Georgics* aptly sum up the quality of Isobel Hutchison's life in her final decade. Abundant good health, stamina, courage, and a measure of good luck had carried her through a long and adventurous life. Now she would need stamina, fortitude, and a sense of humor to endure the hardships of her final years. Property rich but cash poor, crippled and then totally immobilized by arthritis, she could no longer roam even through the extensive grounds of Carlowrie. The freedom she loved had to be sought in books and in letting her mind wander back along those northern trails that few others had traveled. Courage, gentility, and humor would be with her to the end.

In the course of her travels Isobel formed deep and abiding friendships, and the correspondence files in the archives are thick with letters from those she met. Dorthe, the Greenlandic kivfak, maintained contact from 1929 for at least twenty years, despite the language barrier and the sporadic mail service. The friendship with Ruth Reat of Nome went beyond wartime food parcels and plants collected for the herbarium of the Royal Botanic Garden in Edinburgh. The two women had found many common bonds during the month of Isobel's unavoidable stay in Nome in 1933, and these were reinforced on her brief return in 1936. A relative of Reat's called at Carlowrie during the war, and correspondence between the two women continued for the rest of their lives. A similar relationship continued with Mrs. Greist from Barrow, whose knowledge of Eskimo artifacts was so valuable.

Her male traveling companions provide striking examples of Isobel's capacity for friendship through shared experience. Men she traveled with also became lifelong friends and continued to exchange news and to hope that they would travel together again. Ira Rank, owner of the little trading vessel in which she sailed from Nome to Barrow, suggested future trips, and his crew member, Pete Palsson, defined exactly why she was remembered as a good companion. "A number of women have asked me if they could not have done just as well as you—'Why not?'—providing they had the same ability of making use of what they observed, saw and heard, and were not too high tone to travel on the Trader."[1] It was this ability to adapt to any situation without making demands or drawing attention to herself that prompted the letter of appreciation for her Aleutian book from an officer on the Coast Guard ship on which she had been a passenger. "We considered you a member of the mess rather than a 'lady guest' [Isobel's name for herself in *Stepping Stones from Alaska to Asia*] on board."[2]

In a different vein, nearly ten years after their exploration of Denmark, Maynard Williams was still writing enthusiastic letters: "Dear Good Companion, This is the guy who threatened to get drunk in Copenhagen, either with the joys of freedom or the torture of separation. Actually, the happy memories live on. . . . I do hope you will sell some idea to the *National Geographic*. But times have changed. Dr. Grosvenor knew, liked and encouraged us. . . . I'd love to do another IWH story with you as we did in Denmark [in 1947]."[3] Williams, a writer as well as a photographer for *National Geographic*, had been on assignments with other writers whose artistic temperaments and self-importance made them difficult traveling partners. Even the RCMP constable with whom she climbed the mountain outside Dawson one afternoon took the trouble to write of the pleasure it had been and hoped for a chance to repeat it.

What was the magnetic quality in Isobel's personality that attracted friendship? Her joy in the natural world was infectious and readily communicated to those around her. Added to this were her qualities of honesty, openness, and the capacity to accept discomfort or delay without complaint. On the Coast Guard ship and the little boats in the North, she brought a welcome variety to the exclusively male company without posing a threat, being neither a femme fatale nor a predatory female in search of a mate. The men who traveled with her found a genuinely interested, and consequently responsive, companion with whom conversation flowed naturally or silence was not oppressive.

Isobel had been intimidated when she was the only woman in her theology classes at the University of London, but the more she traveled, the more comfortable she became in the company of men whose broad interests were compatible with her own. As her confidence grew, her ease extended beyond travel to academic or scientific gatherings, where she sometimes found herself the only woman present. Never possessed of a fund of small talk, she was happiest among people involved in their own work whose interests she could share, and the broad spectrum extended from scientists at Cambridge to the workmen at Carlowrie. A favorite walk in her late years took her across the fields to the tenant farm, where she liked to sit in a sunny arbor and talk to Peter McGowan as he worked in his greenhouse.[4]

Described by a journalist who interviewed her as "one of the most feminine of women, gentle and charming, warm and friendly in her manner," Isobel had a fearless independence as a traveler that caught people by surprise.[5] Such contrasts were present throughout her life. As a child she enjoyed sports with her older brother as readily as she took pride in doing beautiful needlework. In Greenland the exhilaration of climbing a mountain peak was balanced by performing the duties of a proper hostess in Umanak. At home at Carlowrie, a morning spent helping the gardener cut up fallen trees might be followed by a social afternoon in Edinburgh visiting elderly relatives or a council meeting of the Royal Scottish Geographical Society. The inspiration for her unusual journeys arose out of feminine intuition rather than masculine logic. At the same time, her feminine nature cloaked the resolute determination, usually considered a masculine trait, needed to accomplish her travels, which were physically demanding, difficult, and sometimes dangerous. Psychology applies the term "gender-role transcendence" to those for whom gender roles have no meaning, for whom human qualities outweigh stereotypes.[6]

Susan Brind Morrow, a modern writer whose book *The Names of Things* describes her solo travels in Egypt and the Sudan with Arab men as her guides, noticed a similar split in her own perception of gender. When sitting among strangers and speaking a foreign tongue or going eagerly into potentially dangerous situations, she felt masculine, whereas her feminine side was content to remain at home.[7] In Isobel Hutchison's case the rift would be identified less as male/female than as comrade/lady.

Isobel expressed femininity in her clothing and her manners. She enjoyed choosing clothes of good material, well cut, practical for travel, and not showy. Photographs taken in Greenland or on her long walks show her wearing a tweed or knitted suit with a softening blouse or decorated jersey. In her diary she mentioned buying a new scarf for the royal garden party in Edinburgh, to dress up her good Liberty silk frock bought years earlier. Even in her time of turmoil in London, when she was short of money, she bought a beautiful hat as an outlet for her feelings. In manner, Isobel was a lady through and through—courteous, considerate, refined, gracious. The code of good manners was like a shield in unconventional and potentially dangerous circumstances. In his first words to Isobel, Gus Masik assured her she would be treated like a lady, and neither he nor she broke the code.

There is no evidence in diaries or letters to suggest that Isobel was ever attracted to marriage. Her independent spirit would have found difficulty adjusting to the subordinate position customary for wives of that period or would have been stifled by the overprotectiveness of a loving husband. Equal partnerships were rare, and the benefits of love and companionship would be outweighed by the loss of total freedom. Isobel had friends, a bevy of relatives, a home, and a modest income; she was absorbed in her writing, painting, and gardening; she was self-reliant and accustomed to making her own decisions; and she was free to travel to the ends of the earth. In choosing to remain single she and her sister are part of a largely invisible minority of resilient, resourceful women who lead rich lives unsupported by men.[8]

With total disclosure the current fashion and sexual orientation openly discussed, the question of lesbianism or sexual repression arises. Physical sexuality was a subject that was frequently taboo for discussion by Victorian parents, who believed that knowledge was a threat to modesty or purity.[9] Isobel's innocence of "the facts of life" is noted humorously in her novel when the gardener corrects the young girl who mistakes a cow for a bull. " 'Yon's no a bull ava, miss; yon's a coo!' There was an accent in his voice when it fell upon the word 'coo' which gives you the whole sex problem in a nutshell, if you must have such an unwholesome and unnecessary thing as a sex problem in a world that consists not so much of men and women as of different sorts of individuals."[10] There was more truth than fiction in her novel, and those few words encapsulate Isobel's attitude toward sex. As already quoted from a diary written at Studley College when she was thirty, she felt as though she belonged to neither sex. Sex was not a subject that interested her. After she recovered from her emotional turmoil in 1920, her spiritual nature was nourished by a transcendent inner peace that depended for support on no earthly person, man or woman.

Further confirmation of Isobel's lack of sexuality occurs in a reading of her horoscope in 1936 by an Edinburgh astrologer, Hilda Pagan. Among the characteristics Pagan cited are independence, responsiveness to joyous impulse, creativeness with no wish to have others guide or rule, deploring of idleness, and delight in taking pains. All are recognizable as aspects of Isobel's character and borne out in what we have seen of her work and travels. In addition, Isobel was born under the sign of Gemini, but in her horoscope the twins are very young, with two resulting traits: she is young in spirit, and sex means very little to her.[11] Isobel gave some credence to her horoscope, mentioning years later in an interview that the stars had guided her travel choices.[12]

Isobel acknowledged the kindness she received on her travels with appropriate gifts. Nearly fifty years after the wife of RCMP Commissioner Allard entertained Isobel at Dawson City in the Yukon, Dorothy Allard was still receiving the annual Christmas gift of a Scottish calendar, the final one just weeks before Isobel died.[13] The three women at Shingle Point appreciated the books, specifically chosen for each one and dispatched to the lonely mission in the Far North as soon as Isobel reached Scotland. Expressing the hope that she would visit them again, the writer of the letter showed her admiration for Isobel's brave spirit: "You little know the amount of silent tribute you received from people here who knowing the country could more readily understand a few of the discomforts, and the true courage that many of your lonely experiences called for."[14]

The friendship between Isobel and Gus Masik, the Estonian miner and trader who shared his cabin with her for more than a month, and who took her by dogsled to Herschel Island, was in a special category. His letters, written in phonetic English, are preserved in a separate file and not mixed in with other correspondence, indicating the special place he occupied in her affections. As Isobel had explained to Gus's sister in Estonia through an interpreter, her feeling for Gus was as a sister for a brother. Bound by a love of adventure and the North, their life experiences had little else in common. With his restless nature, Gus roamed the west coast of America from Nome to California, took up stock farming near Seattle, married twice disastrously, went to the mines in Quebec to recoup his fortune after the failure of the second marriage, and finally settled into a retirement home near Seattle. On his death in May 1976, Isobel received a moving letter from Beulah Hermann, the nurse who attended him:

> Gus refused oxygen, wanted nature to take its course. Head nurse stayed all night with him. When asked if everything was alright with his soul he replied: "Yes, Beulah knows." I told her I thought everything had been alright with his soul all his life as he had always been a very special type of man. All the nurses were very fond of him and all wanted to say good-bye to him. I have never seen anyone die with such dignity and peace as our dear Gus. He was conscious to the end, just closed his eyes and stopped breathing.
>
> Your travels together were the happiest time of his life. Your letters were very important to him.

Enclosed in the letter was the pin that Gus always wore, bearing the words "We blaze the trail, civilization follows."[15]

Sometime late in the 1960s the oldest Hutchison sister, Nita, returned to Carlowrie. Though clever and original as an artist and writer when she was young, early in her married life Nita adopted the habit of spending much of her time in bed. As the wife of a paymaster in the Royal Navy with frequent postings, Nita spent her life in rented accommodations in the south of England or abroad. In the summers she often went north in Scotland to Bettyhill, where her sisters would rent a cottage nearby. Knowing of Nita's indolent disposition, it was with reluctance that her two sisters loyally agreed that she could return to live at Carlowrie. After the war it was increasingly difficult to find staff to maintain and run the big house, and more of the work fell to Isobel in particular. True to form, when Nita moved into Carlowrie nothing could induce her to leave her bed, even when painters were redecorating her room. She happily covered herself in a dust sheet, put a towel over her head, and enjoyed the unaccustomed company while they worked around and above her.[16] Nita died in 1971.

Isobel and her younger sister Hilda had been together at Carlowrie since the death of their mother in 1931, each pursuing her own interests. Hilda was the more forthright of the two, and when there were visitors it was usually she who acted as hostess. On those occasions Isobel, with her sweetness of manner, was welcoming but content to leave the dominant role to Hilda. Those who came to call in the late years remembered Isobel's penetrating blue eyes, which often seemed to be looking into the distance, but they heard nothing of her surprising adventures or achievements.[17] When a new family bought the neighboring estate of Foxhall about 1965, the two elderly sisters, properly attired in their tweeds and gloves, walked over to pay a formal call and leave visiting cards. The house at Foxhall was under renovation, the furniture still in packing cases, and tea was served in an upstairs bedroom. It was many years before the flustered hostess learned that Isobel had taken tea in circumstances much less formal, in Greenlanders' huts or trappers' cabins in Alaska.[18] In their walks about the country surrounding Carlowrie or going to church in Kirkliston, Isobel was usually in the lead, being the more energetic of the two. Hilda died in 1976, a few months after Gus Masik, and Isobel was alone at Carlowrie with a housekeeper, a gardener, and a practical nurse.

Gradually immobilized by arthritis, in her last years Isobel was confined to a wheelchair on the main floor of the house. It shows the loyalty she prompted in people that the elderly practical nurse who came to help out for a few days after Isobel had an operation on her knee remained for the rest of Isobel's life, about twelve years.

Isobel had outlived all of her family and most of her friends. The important exception was Medina Lewis, ten years her junior, whom she had met when both were students at Studley College. In the end it was Medina who collected and sorted all of Isobel's papers and added them to the collection already begun in the National Library of Scotland. It was also Medina who persuaded the library to hold an exhibition of papers, paintings, and artifacts to mark her friend's singular accomplishments. Sadly, Medina did not live to see the Isobel Wylie Hutchison Exhibition in the National Library in 1987.

Medina was experienced at dealing with archival material. During the years that she was tied to her home in Newtown as the companion to her widowed mother, she had sorted and preserved the papers of the nineteenth-century socialist reformer Robert Owen, who was born in Newtown in

1771, and she helped to found the Robert Owen Museum there. By 1960, when she was free to travel, competent and energetic Medina, with a talent for organizing and facilitating, was in demand to help aristocratic friends organize and refurbish their neglected estates, from Sligo in Ireland to Lacock Abbey in the southwest of England. It was a natural extension of these experiences for her to take charge of Isobel's estate when the time came. Medina was proud of her friend and wanted to ensure that she was remembered.[19]

The accomplishments of Isobel Hutchison are many and varied. Her books, now out of print, take readers into remote areas of the world and capture both the beauty and the isolation. Written from the perspective of a woman, they emphasize domestic arrangements rather than physical hardships, giving a new twist to northern adventure writing. Not for her the tales of heroic battles with the elements or the terrain that are favored by male writers. Whether she was trekking by dog team in subzero weather, traveling by rowboat in a Greenland fjord, or icebound in Arctic waters, she made it seem easy and natural. Her travels in the Far North took place just before the airplane changed things forever. She was one of the last to experience traditional northern travel and the life of the isolated traders and trappers, cut off from the outside world for more than half the year.

Although she considered herself an amateur, Isobel Hutchison's paintings were often hung in juried exhibitions of the Scottish Women Artists and at least once were shown in the Royal Scottish Academy. Her poetry, particularly when it was her chief form of expression, frequently won prizes in competitions in papers such as the *Westminster Gazette* and the *Bookman*.[20] Her articles for newspapers, journals, and *National Geographic,* along with her lectures and broadcasts, gave her considerable prominence during the 1940s and 1950s.

The plants she collected in little-known corners of the North are held in the herbaria of Kew Gardens, the Museum of Natural History of the British Museum, and the Royal Botanic Garden in Edinburgh. To assess the importance of her work it is necessary to understand something of the purpose of herbaria. They are an accumulation of plants collected over time and filed by genus and species, not by region or collector. One of their several uses is to map the distribution of plants and to plot distribution maps of species in a region, country, continent or worldwide. Individual plant collectors are important only for what they contribute to the total knowledge.

The collections Isobel Hutchison made were extensive, properly labeled, and identified as closely as possible, the final identification being done by experts in the field. In identifying her Greenland plants she sought the help of Dr. Morten Porsild in Greenland, and for her Aleutian plants she was in contact with the acknowledged expert, Dr. Eric Hultén in Sweden. Since her collections were made in the North, they were frequently alpine varieties of common species and not new plants. At Kew the North American material is limited, the main interest being in the tropical and subtropical regions of the world.[21] In the herbaria her mounted specimens are interleaved with those of the same species found by other collectors at different times in roughly the same area.[22]

The artifacts Isobel Hutchison collected are treasured by the Royal Scottish Museum of Edinburgh, the Scott Polar Research Institute, and the Cambridge University Museum of Archaeology and Anthropology, where some are on permanent display in the Arctic exhibit. For Cambridge and Edinburgh, she purchased artifacts with money they provided and had a free hand in carrying out their commissions. The artifacts at the Scott Polar Research Institute were a gift from Isobel's own collection, as are some of those in Edinburgh. She had a discerning eye for quality and readily accepted advice from persons who knew the territory, especially in Alaska. Artifacts are more likely to be unique than plants, and her collections were made at a time and in places where few others were operating, giving them added value.[23]

Working as a true amateur, without professional training or academic qualifications, Isobel achieved her objectives in her own distinctive way. Although her work as a collector of plants and artifacts provided her with credentials that opened doors, no institution sponsored her travel. Figuratively speaking, the occupation of collecting stamped her passport, but she still had to pay for her own tickets. Although she received honors in Britain in recognition of her explorations in Iceland, Greenland, and northern Alaska, it was in Denmark that her travels were most appreciated. In Britain, being both an amateur and a woman tended to diminish her accomplishments in the eyes of the establishment, whereas Danish scientists and explorers recognized her as a colleague and friend. Even the king of Denmark regularly acknowledged her birthday.[24]

As an independent scholar, Isobel Hutchison read widely all her life, constantly exploring unexpected byways that beckoned at every turn. Her interests encompassed world history as well as the Scottish chronicles, geology along with botany, Celtic folklore as much as the spiritual teachings of the great philosophers, including the Indian mystic Rabindranath Tagore. Although she would not hesitate to do something that interested her even though it was usually done only by men, she would never have called herself a feminist—merely an independent *person*. In this she was a modern woman ahead of her time. Observing the bounds of conventional society at home, she was prepared to be as unconventional as necessary in her travels.

Travel and writing defined the person of Isobel Hutchison. Writing came first, but travel done for the pleasure of looking and learning, not from a sense of duty or mission, gave meaning not only to her writing but to her whole life.

APPENDIX I
The Literature of Travel and Adventure

Travelers' tales are as old as time.[1] Since the first hunter-gatherers moved across the land and settled in agrarian communities, there have always been individuals whose curiosity took them out beyond the boundaries of the known world. On their return, they held their listeners spellbound around the campfire with stories of where they had been and what they had seen. Some of the stories of epic journeys have become enshrined in mythology. The wanderings of Ulysses, *The Odyssey,* would become the first great piece of travel writing when transcribed by the poet Homer rather than told by the traveler himself. The classical scholar Joseph Campbell, in *The Hero with a Thousand Faces,* has analyzed the journeys of mythology and found that each follows a pattern beginning with a quest. In simple terms, the hero has first to break his ties with home before he faces a series of challenges and dangers that are thrust into his path as he goes in search of his goal. Finally, after successfully overcoming all the obstacles, he re-enters the world he left behind, only to find himself changed by his experiences.

What men can do, women can also do, but they have done it and continue to do it differently. In the literary tradition of journeys of exploration, men such as Sir Richard Burton or Wilfred Thesiger portray themselves as conquering heroes, overcoming physical nature or prevailing over unfriendly natives to reach geographical objectives that have meaning only within the society from which the travelers came. In general terms, women travel to satisfy curiosity about the unknown, and their writings are more descriptive of places, of people, and of domestic arrangements. Their travel is motivated by a desire for freedom, and when they break the bonds of comfort and convention, they release personalities previously hidden even from themselves.

The nature and extent of the travels of Isobel Hutchison can be appreciated by putting her journeys into context with the adventures of other women of the past and a few of the present time. In her valuable book *Wayward Women,* Jane Robinson provides a compendium of adventurous women over the centuries, with thumbnail biographies of their lives, cross-referencing their categories of travel and including a bibliography of their writings. Mary Russell, in *The Blessings of a Good Thick Skirt,* has expanded on some of the women travelers and their world, with chapters on their variety of purposes, venues, and methods of travel. Mary Morris, a travel writer herself, has edited an anthology, *The Virago Book of Women Travellers,* taking a chronological approach, and Jane Robinson in her more recent *Unsuitable for Ladies* has arranged excerpts of their writings geographically.

The subject of Victorian women travelers has attracted the critical attention of literary scholars. Dea Birkett, in *Spinsters Abroad: Victorian Lady Explorers,* pays particular attention to freedom as a motivating force and finds that Victorian women both were exploited by and exploited the prejudices of their time. In *Imperial Eyes: Travel Writing and Transculturation,* Mary Louise Pratt examines the travel narratives of both men and women to show how the imperialism of science was used by nineteenth-century travelers. She further argues that the civilizing mission and social reformism of women travelers living in a contact zone might be regarded as a form of female imperial intervention. In *Britannia's Daughters* Joanna Trollope looks at travel from the point of view of British women who were dispersed to the outposts of empire for a variety of reasons, only a few as travelers by their own choice, and recognizes that despite initial hardships they benefited from the contact with a wider world.

Analysis of the writings of traveling women visiting the Canadian North has provided material for at least two doctoral dissertations at Canadian universities. Barbara Kelcey, writing "Jingo Belles, Jingo Belles, Dashing through the Snow: White Women and Empire on Canada's Northern Frontier" for the University of Manitoba in 1994, uses the fragmentary history provided by women's collective memoirs to balance the standard male version of northern history and finds that the volume of material is significant enough to add new dimension to the existing history. Lisa LaFramboise in "Travellers in Skirts: Women in English-Language Travel Writing in Canada, 1820–1926," written for the University of Alberta in 1998, argues that women's travel writing negotiates contemporary discourses of femininity and travel to create a specifically feminine travel authority at the intersection of discourses of class, race, and gender. She argues that a variety of strategies are needed to represent the travelers as both adventurers and ladies.

Travel writing can take many forms, from diaries, journals, and letters to polished and literate memoirs. Until recently it was a genre outside the range of study by literary scholars, and in the past decade it has also come under the scrutiny of feminist scholars. Lisa Bloom's *Gender on Ice: American Ideologies of Polar Expeditions* develops the argument that polar exploration and its sponsorship by the National Geographic Society are a statement of masculine dominance. Sherrill Grace's "Gendering Northern Narrative," in *Echoing Silence: Essays on Northern Narrative,* a collection edited by John Moss, examines the ways recent works by Canadian women writers and artists are challenging the aggressively masculine colonialist approach to the North that was previously the predominant position. In her most recent work, " 'A Woman's Way': Canadian Narratives of Northern Discovery," published in *New Worlds: Discovering and Constructing the Unknown in Anglophone Literature,* Grace uses "autobiographics," a feminist theory of women's self-representation, to present a feminist reading of the travel writing of Mina Hubbard and Victoria Jason, whose travels will be discussed later. This is but a sample of the analytical material on women travelers, giving some indication of the breadth of the field and the growing interest in the subject.[2]

The scholarly attention would come as something of a surprise to the women travelers whose writings are noted in this survey. Travel was not something they did to make a point. There are less arduous ways to do so than undertaking journeys to the ends of the earth. Women travelers harbored few doubts about their equality as individuals and were rarely concerned with proving themselves. Far from joining with other women to promote causes, they were too involved in making the most of their own lives. In their manner and their dress they might appear feminine, but they would rarely be classed as feminists.

They were, however, strong individualists, and the reasons they give for breaking out of safe and conventional domestic lives to pursue their dreams are as varied as the women themselves. The purposes for travel can be divided into three general categories: pilgrimage, duty, and pleasure. Pleasure may include the pleasure of studying various sciences including botany, archaeology, and ethnology. Writing about travel often becomes the means of sustaining an addictive habit, but it is secondary to the main purpose. For example, *Full Tilt*, by Dervla Murphy, the account of a solo bicycle trip from Ireland to India, sweeps readers along with its Irish wit and gusto. As she continued to make cycle journeys with catchy titles, however, the writing became a predictable formula. When writing becomes the driving force, it loses the freshness and vitality that are the soul of great travel writing.

Pilgrimage was the earliest form of women's travel. In 1884 a Latin manuscript was discovered in the Biblioteca dell'Accademia Storio-giuridica in Rome in which the Abbess Etheria (or Egeria) recorded the pilgrimage of Saint Silvia of Aquitaine to the Holy Land in A.D. 385. This manuscript aroused such interest that over the next thirty years it was published in a dozen editions in a variety of languages, according to the catalog of the British Museum. Slightly more accessible is the account of the pilgrimage made in 1436 by a privileged Englishwoman, Margery Kempe, which had to be dictated because Kempe could neither read nor write. The Holy Land continued to be a magnet to travelers and tourists alike, and like some other women travelers, Isobel Hutchison chose that destination for her first independent trip abroad.

Other religions have attracted pilgrims as well. The French journalist Alexandra David-Neel spent many years studying and preparing for the journey to Tibet, where she was the first Western woman to enter the sacred city of Lhasa in 1923. Because women were forbidden within the walls, she had to travel in disguise as a Tibetan peasant. *My Journey to Lhasa*, published in 1927, tells a gripping story of escaping detection. A different form of pilgrimage was the bicycle trip made by Bettina Selby from Karachi to Katmandu to observe firsthand something of what she had learned as a student of comparative religion. Her book *Riding the Mountains Down* is a catharsis of the harrowing experiences she endured passing through populous Asian cities, where her carefully chosen cycling clothes did not inhibit sexual predators.

Travel as a form of duty is a broad category, encompassing missionaries, both religious and medical, nurses, teachers, and perhaps the most interesting and diverse group, wives. Missionaries range from the Ursuline nuns who began arriving in the fledgling colony of New France early in the seventeenth century to Mary Slessor, a young Scottish woman who went alone to West Africa as a missionary in 1876. Another remarkable medical missionary was Kate Marsden, who felt called to serve a leper colony in Siberia in 1891 and wrote with black humor of the experience in *On Sledge and Horseback to Outcast Siberian Lepers.*

In the field of nursing, the brilliant example of Florence Nightingale prompted a few untrained middle-class women to escape from Victorian drawing rooms and follow her to the makeshift field hospitals of the Crimea and later to South Africa during the Boer War at the close of the nineteenth century. Thirty years later, nursing had become an established profession that Kate Austen could use to leave home in Australia and go north to the Grenfell Mission at Northwest River in Labrador. There she fell in love with a young American WOP (without pay) teacher and handyman, Elliott Merrick, also at the mission. Kate Austen was not a writer, but her husband convincingly adopted her voice to tell her story in the book *Northern Nurse.*

Teachers who traveled in the course of duty worked mainly on the frontiers of their own countries, apart from a number who traveled to outposts of the British Empire to serve as governesses in the young colonies of Australia and New Zealand. Isobel Hutchison met several such teachers on the north coast of Alaska and in the Yukon, the Aleutians, and the Pribilofs, as well as some who had come from England. *A Schoolteacher in Old Alaska: The Story of Hannah Breece,* edited by Jane Jacobs, is the literate account of the difficulties and hazards faced by a middle-aged single woman during fifteen years of teaching native children in remote communities in southern Alaska and on Kodiak Island, beginning in 1904.

Dutiful wives who traveled include many who accompanied government officials and senior army officers abroad during the glory days of the British Empire in Queen Victoria's reign. Beginning in the 1840s, they were allowed into India along with the wives of civil servants, giving rise to the class of "memsahib," with the vicereine and the "colonel's lady" at the top of the social ladder. It was the 1890s before women accompanied their men to Africa, and their experiences differed from those in India, with fewer servants and greater latitude for freedom.

Commercial enterprises such as the East India Company permitted wives to join their husbands, as did the Hudson's Bay Company in Canada early in the nineteenth century. Their memoirs constitute another branch of this literature. *The Letters of Letitia Hargrave, 1813–54,* describe daily life on the coast of Hudson Bay by the first woman allowed to go so far north with her husband.

Of particular interest to this survey are the writings of wives who accompanied their husbands on extraordinary journeys, sometimes tours of duty, more often of pure adventure. They could range all the way from the dutiful Mary Livingstone, who followed her missionary-explorer husband David Livingstone deep into Africa, giving birth to five children at yearly intervals in the 1840s, to the bizarre case of the adventurous young woman Lucy Irvine who, answering a newspaper advertisement, married a much older man in order to spend a year with him on a desert island in the Torres Strait north of Australia. There are stories to be found from every corner of the globe, but this survey will focus on the North, following the example of Isobel Hutchison.

The Pribilof Islands were one of Isobel's destinations, and *Libby* is the diary of Elizabeth Beaman, who spent a year on those islands in 1879–80. Published by her granddaughter Betty John, the book is filled with graphic descriptions and the charming illustrations of Beaman, a trained artist. Beaman had been granted special permission to accompany her husband through the intervention of the highest authority, President Rutherford Hayes. Though resented by other men on the islands denied the company of their wives, and despite the dangerous winter ice storms, she found beauty in the barren landscape.

Although no official permission was needed for Josephine Diebitsch-Peary to accompany her naval officer husband to the Far North for his first exploration of the Greenland ice cap in 1891, official Washington was shocked. It was thought that no Caucasian woman could possibly live so far north. Josephine Peary's published diary, *My Arctic Journal,* shows her to be spunky, practical, and feminine, capably nursing her husband when he broke his leg at the beginning of the journey, making their temporary Arctic home attractive and cozy, and trekking about their frozen fjord on snowshoes to camp in the snow. All together, Josephine Peary spent three winters and eight summers in the High Arctic, giving birth to their first child in northern Greenland in 1893. She provided practical and psychological support for her husband by raising much-needed funds to rescue his failing expedition in 1895 and was eventually awarded a special gold medal by the National Geographic Society for her contribution to her husband's explorations.

Ella Wallace responded to an oblique proposal of marriage in 1938 to join Thomas Manning on his exploration of the unmapped coast of northern Hudson Bay. The Cambridge University group who had previously traveled with Tom Manning respected him as an experienced northern traveler but one who literally enjoyed discomfort. Tom and Ella Manning spent two and a half years exploring Foxe Basin on his thirty-foot whale-boat, the *Polecat,* with its cranky inboard engine, living in a tent during the summers and in igloos when the *Polecat* was frozen in. Ella Manning wrote about their adventure in *An Igloo for the Night,* with forbearance but without much pleasure. After a later stint together with a survey team in northern Ungava, the Mannings parted company. Tom Manning continued his work as a naturalist in the Arctic, and Ella retired to central Canada with perpetually cold feet from her years in the North.

It was a different sort of honeymoon trip for Margaret (Mardy) Murie in 1924, going up the Koyukuk River into the Endicott Mountains of northern Alaska, where her husband was documenting caribou herds. Olaus Murie had learned from the native guides how to travel comfortably in all conditions. But when Mardy Murie left their first cozy cabin on a frosty morning to run twenty miles a day behind a dog team, she felt pulled in two directions, eager for adventure but reluctant to leave her new role as homemaker. In her memoir *Two in the Far North* she tells of overcoming fear for her husband's safety when she was alone in camp, and she adapted so well to rigorous conditions that he included her on his collecting trips whenever possible, even in the northern Yukon with their young baby.

Of the wives who went north out of duty to their husbands, Mina Hubbard is unusual in that her husband was dead. To honor him, she made her canoe journey in 1905 to explore and map the river system that was thought to flow north to Ungava Bay. Leonidas Hubbard had led an expedition to Labrador in 1903 with two companions, Dillon Wallace and a Métis guide, George Elson. Hubbard had made costly errors in an unforgiving land and died of starvation, while the other two men escaped with their lives. When the widow, Mina, read Wallace's published account of the disastrous journey, *The Lure of the Labrador Wild,* she was driven by grief and fury to set the record straight. She hired George Elson and three other native men and made a triumphant journey to the mouth of the George River, producing a detailed map for the American Geographical Society. What adds even more spice to this story is that she and Dillon Wallace, no longer on speaking terms, set off at the same time from Northwest River in Labrador, the starting point of the 550-mile canoe journey. She achieved the goal of Ungava Bay without incident six weeks ahead of Wallace and wrote *A Woman's Way through Unknown Labrador.* Wallace wrote *The Long Labrador Trail,* and in neither book does one mention the other's existence.

Bluff, hearty, masculine adventure tales, Wallace's books have remained in print since they were written. Hubbard's book went out of print soon after it appeared and has only recently been reissued owing to the growing interest in women travelers. A woman who adopts the masculine "hero" pose to tell her story will not be believed, and publishers may not accept her work. On the other hand, if she writes from a woman's perspective, emphasizing the beauty of the landscape, meeting native people, and the small details of everyday life, the element of "adventure" is missing and her story is dismissed as weak. Sara Mills, in *Discourses of Difference,* discusses the textual strategies and conventions used by female travel writers that set them apart from their male counterparts. Using the model developed by the French theorist Michel Foucault, Mills analyzes the constraints on the reception of women's travel writing and argues that their writings should not be judged against present feminist standards unless there is an awareness of the discursive frameworks involved.

Each of the foregoing "duty" trips was driven by a sense of purpose, without which none of these women would have been tempted to break the ties that bound them to home. In some cases the women were supported by husbands or institutions or both, destinations were chosen for them, and traveling expenses were covered. With all generalizations there are exceptions. Bettina Selby, the pilgrim on her bicycle, and Mina Hubbard, alone with the four aboriginal guides and covering her own expenses, are both anomalies, but both were driven by a sense of duty or purpose. Despite her feeling of freedom in the northern wilderness and her satisfaction in the outcome, Hubbard did not make another journey, whereas Bettina Selby so enjoyed the freedom of travel that she continues to the present day. She rode the length of the Nile to its source deep inside Uganda and wrote *Riding the Desert Trail* (1988), then followed the pilgrim route from France to Santiago de Compostela in Spain, described in *Pilgrim's Road* (1994). The joy of the open road, together with the opportunities the humble bicycle affords her of meeting the ordinary people of the countries she visits, give Selby a place in the third category of travelers, those who travel for their own pleasure.

Among women who traveled for their own pleasure are two famous Victorians, Isabella Bird Bishop and Mary Kingsley, about whom many studies have already been written. A sea voyage was prescribed for Isabella Bird when conventional medical treatment failed to alleviate a spinal condition. From that moment, Bird was free from pain as long as she was on the move, whether on horseback in the Rockies or climbing a volcano in the Sandwich Islands. Each time she returned to her devoted sister in Edinburgh to lead a conventional life, the pain returned. Even marriage to Dr. John Bishop did nothing to impede her progress from one unexplored country to another. He accepted her long absences with good grace, knowing his only formidable rival was the high tableland of central Asia. Between voyages Bishop wrote nine books of travel, and after her penetration into China she was invited to speak to the Royal Geographical Society. She was then admitted to membership in that sanctum of male travelers, creating a furor among the more conservative members. *A Curious Life for a Lady,* by Pat Barr, tells the story of Bishop's life and travels, with many excerpts from her books.

In contrast to Bishop, Mary Kingsley made only two journeys, both to West Africa in the 1890s, where death from tropical diseases was the predicted outcome. Until age thirty-one Kingsley's life, as the unmarried daughter at home, was one of continuous service to her family. When both her parents died within six weeks of each other and her brother left for China, she was suddenly free of obligations. Her father, who had traveled

all his life as a naturalist-explorer, had depended on Mary to translate
and organize the material he brought back from abroad, even though she
was never given an education. From her father's eclectic library she read
voraciously and educated herself. Kingsley's journey took her up the Ogowe
River in what was then the French Congo and is now Gabon. She traveled
part of the time with the people of the Fan tribe, whom their neighbors
feared as ferocious warriors and cannibals. Maintaining English decorum
and wearing proper Victorian dress, Kingsley had a sense of humor and
forthright manner that carried her through daunting situations while she
collected specimens and at the same time carried on trade to pay her way.
She recorded the customs of the different tribes she met, paying special
attention to their fetishes, and though she was not trained as a naturalist or
an anthropologist, her second book, *West African Studies,* expanded on the
vast array of scientific topics that she had merely touched on in *Travels in
West Africa.* Her love affair with Africa took her back as a nurse in the Boer
War, where she caught enteric fever and died at age thirty-eight, mourned by
all who knew her, a brilliant star in the firmament of Victorian travelers.

Although their travel was for pleasure, it suited Victorian ladies to
attach a worthy purpose to their rambles. Where Mary Kingsley pursued
a scientific interest and Isabella Bishop sought to inform the public through
her books, Marianne North used botanical painting as her excuse for continuous travel to all corners of the globe from 1870 to 1885. North's family
included a British prime minister and the master of a Cambridge college.
Her father, a member of Parliament, was her greatest friend and companion,
and until his death in 1869 the two of them spent the summers in southern
Europe and the Mediterranean ports of the Middle East. Once Marianne
North had recovered from the loss of her father, she began to travel abroad,
first to North America and Jamaica, then across the Pacific, and finally to
Chile, visiting every country of botanical interest. She had letters of introduction to many government houses, but she fared equally well in modest
accommodations and was game to tackle any form of transportation to
see a new flower and capture it in paint. Her eight hundred paintings of
lush tropical flowers and trees cover the walls of a specially built gallery
that she donated to Kew Gardens. The diaries of her travels were published
after she died in 1893, under the title she had chosen: *Recollections of a
Happy Life.*

A less well-known Victorian is Gertrude Bell, who was raised by her stepmother in a wealthy home and studied history at Oxford. On a holiday trip to Teheran, she fell in love with a penniless diplomat and felt the full weight of Victorian parental authority when her father forbade their marriage. Bell reacted by immersing herself in study of Arabic and Persian and turned to archaeology in Syria and Asia Minor. Over the next dozen years she lived continuously in the desert, traveling with a full complement of servants and a train of twenty camels. By the outbreak of the 1914–18 war, her knowledge of the region was so extensive that the British government invited her to become a member of the Intelligence Corps of Arabia. Of a naturally melancholy disposition and twice unable to marry men she loved, in 1926 she took her own life in the desert. She wrote *The Desert and the Sown,* and her letters were published in two volumes after her death. Her biography by Janet Wallach, *Desert Queen: The Extraordinary Life of Gertrude Bell, Adventurer, Adviser to Kings, Ally to Lawrence of Arabia,* further raised her profile.

A recent biography by Jane Geniesse, *Passionate Nomad: The Life of Freya Stark,* has reintroduced this scholar of Arabian language and life to the reading public. Freya Stark, born at the end of Queen Victoria's reign, was raised in the Edwardian era and was only a little younger than Isobel Hutchison, with whom she shares a similar background. Both were born into wealthy families, although Stark's family was slightly higher up the social scale, being landed gentry with large holdings in Devon. Her homelife was disrupted when her mother took Freya and her sister to live in Italy, and after the rift they saw their much loved father only on rare occasions. In the preface to her first and best-known travel book, *The Valley of the Assassins,* she admitted that she traveled purely for the pleasure of it, the same reason she had learned Arabic and Persian. Her thorough knowledge of prewar Iran and Iraq gave her a useful role as a propagandist in the 1939–45 war, with postings in Yemen, Cairo, and Baghdad and even a tour of the United States, described in *Dust in the Lion's Paw.*

It often seems as though British Victorian women had the monopoly on adventurous travel, and there is a reason why they are predominant. Well-to-do American women at the same time were making the obligatory cultural tour of European capitals. The growth of the British Empire made heavy demands on manpower to maintain the links that had been forged around the world, creating a rising tide of single women stranded at home. The life of the middle-class spinster, without prospects of marriage, was not an enviable one. With no opportunity to earn a living except as a governess, she was often a slave to her family. Young women with spirit sought escape in any form, including (as we have seen in the section on travel as a duty) emigrating to the colonies as governesses or taking any form of work in the hope of finding husbands. Those fortunate enough to have a small private income were able to indulge their passion for travel. Within the confines of the family structure, they were like coiled springs waiting to be released. When release came, often through the death of an ailing parent, they shot off in surprising directions.

Women from other nations had the same longings. Mrs. Ida Pfeiffer, born in Vienna in 1797, the only girl in a wealthy family of six children, was brought up wearing boys' clothing and competing in her brothers' games. After her father died, her mother attempted to direct her toward marriage but refused to let her marry the one person who interested her. To please her mother, she married an elderly lawyer who then lost his fortune by holding fast to his principles in a government scandal. Ida had always dreamed of travel, and as soon as her sons were educated and an inheritance from her mother released her from poverty, she made her pilgrimage to Jerusalem. Next she set about learning English and Danish and the art of making daguerreotypes, in preparation for a visit to Iceland. She wrote books about the two journeys that began to earn money for her. The floodgate had opened, and the world was her oyster. She left in 1846 for a trip around the world, arriving back two years later to publish another book. She was insatiable; a second world tour followed in 1851, this time taking three years. Staying home long enough to visit her husband and sons and see her latest book through the press, she embarked on another voyage in 1861, dying en route in Madagascar. Described by her son as short, thin, and slightly bent, she had the appearance and manners of a practical housewife. With strength of character, obstinacy, and personal courage, indifferent to pain and the ordinary conveniences of life, she was driven by the desire to add something to the stock of human knowledge. *The Last Travels of Ida Pfeiffer* was published by her son and includes an autobiographical memoir.

Apart from Pfeiffer's Icelandic trip, there are few instances of women going north for pleasure. Greenland was not open to tourists, and only travelers with a valid scientific purpose could get permission to visit; hence Isobel Hutchison's discovery of a vocation as a plant collector. In northern Canada, transportation was controlled by the Hudson's Bay Company. When the Canadian Pacific Railway opened a line to Edmonton in 1891, it became possible for a determined traveler to get almost to the Arctic coast and return during a single summer season. Nevertheless, it was a rough trip by wagon train, fur brigade scows, and finally paddle-wheel steamer, and it took stamina.

Only a few women made this journey for pleasure and wrote accounts of it. Elizabeth Taylor, the daughter of the American consul in Winnipeg, was the first of these to reach Fort McPherson on the Peel River, the northern limit for the steamer in 1892. *The Far Islands and Other Cold Places,* a collection of magazine articles about her travels, has recently been published. Another American, Emma Colcleugh, making the same journey in 1894, published her account in 1932 as a reminiscence in the Providence, Rhode Island, *Evening Bulletin.* The route had not changed when Agnes Deans Cameron and her niece, Jessie Cameron, followed it in 1908, but in her imagination Agnes Cameron continued down the Mackenzie delta "to spend time among the Eskimo." *The New North,* informative and colorful, explores the past and present of the region in an opinionated and gossipy fashion. For the journey, Elizabeth Taylor had a stylish traveling costume made in Paris, and the Camerons both wore skirts that swept the ground, topped by jackets and hats resembling those worn by the mounted police. Yet Agnes Cameron posing for the camera holding a rifle, her foot on a dead moose, belies any thought of women travelers as delicate flowers.

The other area for northern travel was the Klondike, where the gold rush attracted a surprising number of women. A few of them, such as Martha Black and Laura Berton, stayed and established homes there and wrote about the experience.[3] A handful of women traveled to the North merely for pleasure and novelty, despite reports of terrible hardships experienced by the hordes of miners streaming into the Yukon beginning in 1897. Lulu Craig, a young schoolteacher, made the trip across the Chilkoot Pass with her brother and his wife and child in March 1898 and stayed until the following summer, writing about it with pleasure in *Glimpses of Sunshine and Shade in the Far North.*

Craig was under the protection of a male relative, unlike two women who sailed from San Francisco in the summer of 1899. Mary Hitchcock, the widow of a naval officer, and Edith Van Buren, grandniece of President Martin Van Buren and the daughter of the American ambassador to Japan, were no ordinary tourists. They traveled with a menagerie as well as many luxury items. The preface to Hitchcock's book, *Two Women in the Klondike,* written by a male friend, emphasizes the heroic nature of their travel with the statement that "the narrative is a tribute to their perseverance and determination and the character of intelligent, fearless Anglo-Saxon women who, among all sorts and conditions of men, never fail to secure protection and respect." In fact, the greatest problem the ladies faced was tedium as they waited more than two weeks at the port of St. Michael for the steamer up the Yukon River. When they at last reached Dawson, they had their enormous tent (forty feet by seventy) erected, hired a man to cook and carry, and rapidly became bright lights of Dawson society. They found the miners entirely courteous, helpful, and trustworthy and enjoyed every facet of Dawson life. As cold weather approached they so dreaded the boredom of the boat trip down the Yukon that they went upriver to Whitehorse and hiked out over the snow-covered White Pass to the railway.

The women who went to the Klondike out of interest were Americans, traveling on their own continent and raised in a society less restrictive than Victorian Britain. Clara Rogers and Gwen Dorrien Smith, two Englishwomen who went north in 1926, knew firsthand the constraints of Victorian propriety. The extension of the Alberta railway during the 1920s had eliminated the wagon train and the fur brigade scows from the northern journey made by their predecessors. Rogers and Dorrien Smith aimed to reach the west coast of Canada by crossing the continental divide at its lowest point. This involved a trip up the Rat River to the height of land and continuing downstream on the Porcupine River to Fort Yukon, where they would find conventional steamers on the Yukon River, then going onward to Vancouver. They hired two Gwich'in guides to take them up the short, turbulent Rat River by canoe, the men wading and lining the boat upstream while the women mainly walked along the rough, tussocky edge of the stream. When leaving England, they had expected to find their next guides to take them downstream at a fur-trading post, long since abandoned. Although they had never canoed before, they not only managed but enjoyed the complete freedom of the wilderness for three days from the time their guides turned back until they reached the next post. Forty years later Clara, by then Lady Vyvyan, wrote the story of their travel in *Arctic Adventure,* republished as *The Ladies, the Gwich'in, and the Rat,* edited by Ian MacLaren and Lisa LaFramboise.

The British women had been preceded in 1925 by an American, Mrs. Laurie Frazeur, who was determined to see as much of the Arctic as possible in a summer season. Arriving at the new village of Aklavik, the last point of call for the Hudson's Bay boat, she hired a motorboat and went farther down the Mackenzie River and along the coast westward to Shingle Point, where she could view the Arctic Ocean. Returning up the Mackenzie to Fort McPherson, in spite of the gloomy predictions of officials Frazeur hired a guide to take her up the Rat River. After days of rain with the river steadily rising, she turned back to Fort McPherson, hired a new guide with pack dogs, and made the six-day trek across the land trail, usually used only in winter. She found a boat going down the Porcupine River to Fort Yukon, where she chartered a plane and flew to Fairbanks. The only account of this highly adventurous trip by a determined woman traveling alone was "From the Mackenzie to the Yukon," published on December 15, 1925, in a small Seattle journal, the *Mountaineer.* Frazeur, a college professor and an alpinist who had climbed in Greece and in the Canadian Rockies, was not one to be deterred by officialdom. Her sense of knowing what she wanted to see and her determination to carry out her plans is in the spirit of Isobel Hutchison, whose trip around the north coast of Alaska remains unique.

There are three distinguishing features to the journeys of Isobel Hutchison: she traveled alone, she traveled in the North, and she traveled to collect plants. Women, traveling alone, continue to make journeys that involve uncertainty, discomfort, and danger, not only in the North but wherever they choose to go. In 1977 Robyn Davidson, a young Australian, made a solo camel trek across the desert center of her country from Alice Springs to Perth, writing of it in *Tracks,* and after a recent camel journey with nomadic tribes in India she produced *Desert Places.* In the past decade an American scholar of etymology, Susan Brind Morrow, has roamed the Egyptian desert and written a luminous meditation on its austere splendor in *The Names of Things.* A Canadian, Victoria Jason, already a grandmother at fifty, traversed the Northwest Passage by kayak over the course of four summers, two of them entirely alone. *Kabloona in the Yellow Kayak* tells the story. Helen Thayer, an American athlete, having made the trip once with her husband, skied to the magnetic pole with only her dog for company and wrote *Polar Dream*. Margaret Mee, a British artist living in South America, made many journeys on the tributaries of the Amazon River, with a sense of urgency to record the plants that are disappearing as the rain forests are cleared for planting. Her life's work is *In Search of Flowers of the Amazon Forest*.

Some of these journeys were precipitated by tragic circumstances that had torn apart the fabric of safe, comfortable lives. Some were dictated by the need to discover the unknown, and some sprang from the sense of adventure and enjoyment. Isobel Hutchison's travel had its roots in all of those reasons. Wealth is not a deciding factor. The travel undertaken is rarely luxurious, and all that is needed is enough money to permit independent movement. Desire and determination are the other essential ingredients. Whatever the purpose, women travelers make it clear in their writings that cutting the ties to home and convention and striking out for distant horizons is the path of freedom. Coming from the British tradition of a class-bound society, Hutchison felt that strong sense of freedom when she was in the North to an even greater extent than the American and Canadian travelers. Although she was only one in a galaxy of women travelers, her star shines bright and solitary in the Northern Hemisphere.

APPENDIX 2

Hutchison and Wylie Family Trees

Hutchison

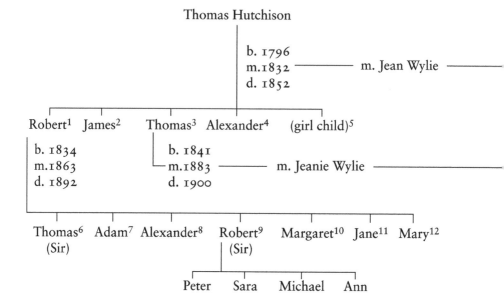

Thomas Hutchison

b. 1796
m.1832 ——————— m. Jean Wylie ——————
d. 1852

Robert[1] James[2] Thomas[3] Alexander[4] (girl child)[5]

Robert[1]
b. 1834
m.1863
d. 1892

Thomas[3]
b. 1841
m.1883 ——————— m. Jeanie Wylie ——————
d. 1900

Thomas[6] Adam[7] Alexander[8] Robert[9] Margaret[10] Jane[11] Mary[12]
(Sir)   (Sir)

Peter   Sara   Michael   Ann

NOTES

1. Born 1834, married Mary Jemima Tait, daughter of Rev. Adam Duncan Tait, died 1892.
2. Married Isabelle Field of Moreland; no issue.
3. Married Jeanie Wylie
4. Died unmarried.
5. Died young, never at Carlowrie.
6. Born 1886, Lord Provost of Edinburgh, married Jane, daughter of Alex Oglivy Spence, died 1925.
7. (Rev.) born 1868, married Margaret, daughter of Andrew Menzies of Babinock.
8. Born 1870, unmarried, died 1898.
9. Born 1871, M.D., FRCP, Bt., married Laetitia Norah, daughter of Rev. Canon Ede, died 1960.
10. Married 1886, Rev. Robert Fisher, died 1898.
11. Married Rev. Harcourt Davidson, died 1926.
12. Married Charles W. A. Taitson.

Wylie

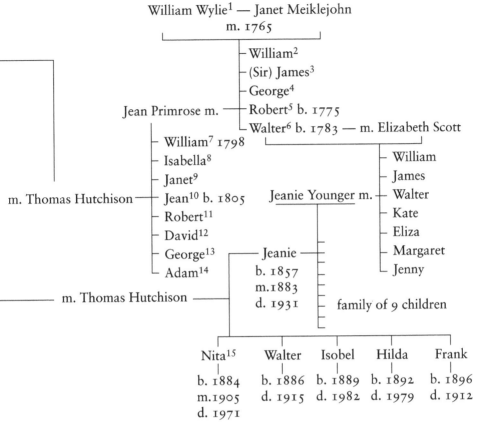

William Wylie[1] — Janet Meiklejohn
m. 1765

- William[2]
- (Sir) James[3]
- George[4]

Jean Primrose m. — Robert[5] b. 1775

- Walter[6] b. 1783 — m. Elizabeth Scott

- William[7] 1798
- Isabella[8]
- Janet[9]

m. Thomas Hutchison — Jean[10] b. 1805   Jeanie Younger m.

- William
- James
- Walter

- Robert[11]
- David[12]
- George[13]
- Adam[14]

- Kate
- Eliza
- Margaret
- Jenny

— Jeanie —
b. 1857
m. Thomas Hutchison —   m.1883
d. 1931   family of 9 children

Nita[15]   Walter   Isobel   Hilda   Frank

b. 1884   b. 1886   b. 1889   b. 1892   b. 1896
m.1905   d. 1915   d. 1982   d. 1979   d. 1912
d. 1971

NOTES

1. Farmer at Kincardine.
2. Born 1766, married Ann Stuport, died 1827.
3. Born 1768, surgeon to czars, baronet; unmarried, died 1854.
4. Born 1771, married Mary Muir, died 1824.
5. Born 1775, married Jean Primrose.
6. Born 1783, married Elizabeth Scott (said to be a runaway marriage, the bride climbing down the ivy), died 1871.
7. Born 1798, lived at Leith.
8. Born 1800, spinster, died 1881.
9. Born 1802, married 1830, D. Morrison of Montreal.
10. Born 1805, married Thomas Hutchison; Carlowrie built in 1852.
11. Born 1808, married 1834, Mathilda Bogart, Annapolis, N.S., Canada; died 1836 on his ship *Edward Thorne*.
12. Born 1810, died 1836 in Moscow.
13. Born 1813, died 1848, drowned at sea.
14. Born 1814, died 1834, drowned at sea.
15. married naval paymaster Padwick, no issue.

Source: Courtesy of Sir Peter Hutchison (abridged).

APPENDIX 3

Extract from Appendix of
North to the Rime-Ringed Sun

List of plants collected in Alaska and Yukon by the author from June to August 1933 for the Royal Herbarium of Kew, the numbers being those of the Herbarium. The species are arranged according to Bentham and Hooker, with extra-British genera at the end of their families.

IV. CRUCIFERAE

Barbarea sp. cf. *B. orthoceras* Ledeb. Nome, Alaska. 16th July. Altitude about 500 feet. Habitat, dry roadsides, &c. Common. (504.)
Arabis ambigua DC. (But fruit required for definite determination.) Dawson, Yukon. 23rd June. Altitude 1080 feet. Habitat, open hillside. Frequent. (180.)
Arabis lyrata L. (But no fruit.) Mendenhall glacier, Juneau, Alaska. 13th June. Altitude near sea level. Habitat, moist pasture. Growing on wet ground under glacier. (38, 39.)
Cardamine sp., possibly *C. Blaisdellii Eastwood*. Nome, Alaska. 3rd July. Altitude 200–300 feet. Habitat, open hillsides. Not common. (290.)
Cardamine digitata Richards. Singnat shelter cabin, about 25 miles south of Barrow village. August. Altitude sea level. Occasional on wet tundra. (630.)
Cardamine purpurea Ch. et Schl. Nome, Alaska. 5th July. Altitude about 200 feet. Habitat, hillsides. Common on open tundra. (322–324.)
Hesperis pallasii (Pursh) Torr. et Gr. Point Hope, Alaska. July. (From the collection made by the schoolmaster, Mr. Messenger.) Frequent. (688–689.)

Sisymbrium sophioides Fisch. Nenana, Alaska. 26th June. Habitat, dry meadows, &c. Fairly common. (188, 189.)

Cochlearia fenestrata R. Br. 25 miles south of Barrow. 20th August. Common on shore. (625.)

Erysimum asperum (Nutt.) DC., sens. lat. Dawson, Yukon. 23rd June. Altitude about 1080 feet. Not very frequent. On hill behind St. Mary's Hospital. (116–120.)

Draba alpina L. var. *algida* (Adams) Regel. 25 miles south of Barrow. August. Altitude near sea level. Occasional on tundra. Also at Cape Prince of Wales. (637–638, 677.)

Draba borealis DC. Nome, Alaska. Cape Prince of Wales, Alaska. July–August. Altitude 300 feet at Nome. Habitat, hillsides. Fairly common. (292–294, 662.)

Draba sp.? (Material too scanty and no fruit.) Nome, Alaska. 10th July. Altitude near sea level. Fairly frequent. Dry roadside. (367, 368.)

Parrya macrocarpa R. Br. Nome, Alaska. 3rd July. Altitude 200–300 feet. Habitat, hills and moorland. Very common on tundra all round Nome. (285–286–289, 686–687.)

Parrya? ("I believe a form of *P. macrocarpa* R. Br.") Bear Creek, Dawson. 22nd June. Altitude 1200 feet. Near Klondike river-bank. (148–151.)

Notes

INTRODUCTION

1. Cited in a letter to the author from the keeper of manuscripts, Norman Reid, University of St. Andrews, September 1, 1999.
2. Trust deed signed by Thomas Hutchison, April 20, 1900.
3. Leonard Huxley, quoted in Raby, *Bright Paradise,* 29.
4. In addition to sending collections of plants to Kew, the Canadian pioneer Catherine Parr Traill published *Canadian Wild Flowers.* The difficulties of an early settler in Australia and her pleasure in the contact that plant collecting provided with the world she had left behind are described in Hasluck, *Portrait with Background.*
5. The membership of such groups was almost entirely male, and the British university participants belonged to an exclusive club. The rare female exceptions were the wives of two British scientists who wintered with their husbands in eastern Greenland in 1936 and a wealthy American, Louise Boyd, who organized, financed, and led six voyages to the Greenland Sea between 1931 and 1938. A retrospective on the Scott Polar Research Institute written in 1945 by the founder, Frank Debenham, and a survey of the *Polar Record,* volume 2, show that Danes, Dutch, Norwegians, Finns, French, Polish, Germans, Italians, Russians, and Swedes were involved in Arctic expeditions, and among the British and Americans there were groups from Cambridge, Imperial College, Merton College Oxford, Oxford University, St. Andrews, and the University of Michigan. In addition, in 1935–38 Captain Bob Bartlett of polar fame sailed north annually for various American scientific institutions, taking along university students who worked as deck hands at their own expense.
6. Hoyle, "Women of Determination," 125–29.

1. CARLOWRIE

1. Much of the information about the Hutchison family comes from an unpublished paper by Sir Peter Hutchison, "The Hutchisons of Carlowrie: A Sketch of a Lothian family, 1852–1982." Some of the details of this material were in a letter to Sir Peter from Mr. Makey, archivist of the Edinburgh Corporation, July 10, 1968.
2. Letter from Isobel Hutchison to Peter Hutchison, January 26, 1969, from personal papers of Sir Peter Hutchison. Sir James Wylie's life is covered in the *Scottish Dictionary of National Biography.*

3. During Thomas Hutchison's term as provost of Leith, Queen Victoria visited the town. The chair in which the diminutive monarch was seated for the occasion is the only relic of Thomas Hutchison's remaining in the grand central entrance hall of Carlowrie.

4. Sir Peter's paper includes a long description of Carlowrie quoted from a book on Lothian architecture by Colin MacWilliam. In addition, I visited Carlowrie on two occasions, April 29, 1997, and October 15, 1998.

5. The life of the youngest son of Robert Hutchison, Sir Robert, father of Sir Peter, is in the *Scottish Dictionary of National Biography*.

6. Diaries of IWH, National Library of Scotland (hereafter NLS), acc. 9713, nos. 4 and 5.

7. Hutchison, *Original Companions*, 52–53, 54.

8. Archivist, Edinburgh Corporation, private papers of Sir Peter Hutchison. The younger Thomas Hutchison became lord provost of Edinburgh in 1921 and was knighted in 1923.

9. NLS, acc. 5509, no. 3, letter to IWH from Sir John O. Miller, 1938.

10. Notes on Jeanie Hutchison née Wylie, handwritten by Medina Lewis from personal papers of Sir Peter Hutchison.

11. Dyhouse, *Feminism and the Family in England*, 12.

12. NLS, acc. 9713, no. 87, containing twenty-three issues of the *Scribbler*, a magazine for family and friends, 1903–11.

13. NLS, acc. 9713, no. 5, IWH diary 1904.

14. Newspaper account of the closing of Rothesay House, "62- Year-Old Edinburgh Girls' School to Close," *Edinburgh Evening Dispatch*, July 22, 1957, 32.

15. Jane McDermid, "Women and Education," in *Women's History: Britain, 1850–1945*, edited by Jane Purvis (London: University College London Press, 1995), 112–17.

16. Notes by Medina Lewis, personal papers, Sir Peter Hutchison.

17. Conversation with Nessie McGowan, widow of the tenant farmer at Wheatlands Farm, April 30, 1997.

18. Conversation with Sir Peter Hutchison, May 5, 1997.

19. Naval Manning Agency, Portsmouth, Statement of Royal Navy service of Victor Octavius Padwick, January 21, 1999; other details of Nita's life were gleaned from the diaries and journals of IWH.

20. Dyhouse, *Feminism and the Family in England*, 9–27.

21. Crow, *Edwardian Woman*, 140; Bentley, *Edwardian Album*.

2. SEARCH FOR MEANING

1. Hutchison, *Original Companions*, 30 and 92.

2. NLS, acc. 8138, box 9, no. 1, "Literary Venture" notebooks, 1910–44; acc. 9713, no. 74, Record of earnings from writings, 1911–63.

3. Ibid.

4. NLS, acc. 8138, box 9, no. 3, diary and articles, 1914–40.

5. Gail Braybon, "The Need for Women's Labour in the First World War," in *The*

Changing Experience of Women, edited by Elizabeth Whitelegg (Oxford: Basil Blackwell, 1982), 91–93.

6. Conversation with Katrina Burnett, niece of Medina Lewis, October 21–22, 1998.

7. Hutchison, *Original Companions,* 1.

8. NLS, acc. 8138, box 9, no. 3, diary and articles, 1914–40, and Hutchison, *Original Companions,* 108–20.

9. From "Hutchisons of Carlowrie" (see above, chap. 1, n. 1).

10. Hutchison, *How Joy Was Found,* reviewed in *Field,* October 20, 1917, in *Glasgow Herald,* November 8, 1917, and in *Times Literary Supplement,* November 15, 1917, 549.

11. Sanecki, *Short History of Studley College.*

12. Doris Bowes Lyons was a granddaughter of the thirteenth earl of Strathmore and a cousin of Elizabeth Bowes Lyons, the mother of Queen Elizabeth II. This is an indication of the social standing of students who attended Studley College.

13. NLS, acc. 9713, no. 14, diary 1917–18.

14. Lewis, *Eveline.*

15. Vicinus, *Independent Women,* 121–62, covers the history of university education for women in England.

16. Lewis, *Dew on the Grass.*

17. Letter from Katrina Burnett to the author, June 11, 1997.

18. NLS, acc. 8138, box 1, no. 2, diary 1919.

19. Vicinus, *Independent Women,* 165; in addition, chaps. 4 and 5 have much to say on the subject of "homoerotic friendship."

20. NLS, acc. 8138, box 1, no. 2, diary 1919.

21. Letter to the author from the Office of the Academic Registrar, King's College, London, November 6, 1997.

22. NLS, acc. 8138, box 1, no. 3, diary 1920. The period in London is covered in detail in the second part of Hutchison, *Original Companions,* beginning on 183.

23. "Kilmeny" is "The Thirteenth Bard's Song," the central ballad in *The Queen's Wake,* by the nineteenth-century Scottish poet James Hogg.

24. This is described in Hutchison, *Original Companions,* 257–59.

25. NLS, acc. 9713, no. 15, diary 1921.

26. Katrina Burnett believes her grandmother would have had a powerful influence on Isobel Hutchison. Conversation with the author, October 21–22, 1998, and letter, October 22, 1998.

27. NLS, acc. 5509, no. 1, Correspondence, 1913–29, contains the 1923 letters from college friends Campion and Medina "Ginger" Lewis and from residents of Inverness Terrace: de Angioli, Bulkley, and Fidler.

3. MAKING OF A TRAVELER

1. Fussell, *Abroad,* 37–50.

2. Young, *Tourism,* 21.

3. Turner, *Golden Hordes,* 29–60; Fussell, *Norton Book of Travel,* 272–77.

4. NLS, acc. 9713, no. 16, diary 1923.

5. *Times* (London), June 14, 1921, "Our Jerusalem correspondent reports that the body of Miss Lomax, curatrix of the 'Garden Tomb' at Jerusalem, who has been missing since May 30, and for whom search parties had been sent out, was found four days later in the well of her own garden over which the cover had been placed. There was no evidence on the body of violence; but there were traces of robbery in the house; it is feared that Miss Lomax resisted an attempt at burglary, and had been murdered."

6. NLS, acc. 9713, no. 16, diary 1923; letter to the author from Rev. John Murrie, Broxburn, Scotland, January 1999, with information about Dr. Norman MacLean's life.

7. Marian Keith (pseud. for Mary Esther Miller MacGregor), *Under the Grey Olives.* This novel based on a conducted tour of the Holy Land gives a contemporary description of tourists in Jerusalem.

8. Grenfell, *Labrador Looks at the Orient.*

9. NLS, acc. 8138, no. 5, February–April, diary 1924.

10. Personal reminiscence of May Anderson, niece of a maid at Carlowrie during World War I.

11. Sutherland, *Hebridean Journey,* 195.

12. Even Samuel Johnson treated the subject seriously in his *Journey to the Western Islands.*

13. NLS, acc. 9713, no. 74, Record of earnings from writings.

14. Fussell, *Abroad,* 3–15.

15. Morris, *Icelandic Journals.*

16. Descriptions of Iceland can be found in *Iceland,* written for Naval Intelligence at Cambridge, 1942, and Kidson, *Iceland in a Nutshell.*

17. Charcot, *Voyage of the "Why Not?" in the Antarctic,* i–iii.

18. NLS, acc. 8138, box 1, no. 7, July–September, diary 1925, Iceland.

19. Morris, *Icelandic Journals,* 121.

20. Auden and MacNiece, *Letters from Iceland,* 37.

21. Morris, *Icelandic Journals,* 26.

22. *American-Scandinavian Review* 22, no. 2 (1934): 142–52.

4. THROUGH THE ICE BELT

1. A. E. Porsild, "Greenland at the Crossroad," *Arctic* 1, no. 1 (1948): 53–57.

2. NLS, acc. 9713, no. 19, diary 1926; acc. 8138, box 9, no. 1, "Literary Venture" notebook; Nansen, *First Crossing of Greenland;* Graham, *Tramping with a Poet in the Rockies* (the poet is Vachel Lindsay); Banks, *Wild Geese.*

3. *Times Literary Supplement,* April 22, 1926, 304; *Drama Monthly,* Book List, April 1926; *Poetry,* October–December 1926; *Yorkshire Herald,* April 13, 1927.

4. "Lovely Lofoten," *American-Scandinavian Review* 23, no. 1 (1935): 47–57.

5. NLS, acc. 4775, Correspondence 1927–66, relating to Greenland and the Arctic.

6. NLS, acc. 9713, no. 20, diary 1927.

7. NLS, acc. 9713, no. 20, diary 1927.

8. Ejnar Mikkelsen, "East Greenland," in *Greenland*, edited by Kristjan Bure (Copenhagen: Royal Danish Ministry for Foreign Affairs, 1961).

9. Hutchison, *On Greenland's Closed Shore*, 23.

10. Hutchison, *On Greenland's Closed Shore*, 32.

11. Hutchison, *On Greenland's Closed Shore*, 38; NLS, acc. 8138, box 1, no. 8, July–November, diary 1927. In her notebook Isobel uses the more pejorative word "stench."

12. Nansen, *First Crossing of Greenland*, 165–66.

13. Hutchison, *On Greenland's Closed Shore*, 39.

14. Hutchison, *On Greenland's Closed Shore*, 65.

15. This kayak is part of the collection at the National Museums of Scotland in Edinburgh.

16. Augustine Courtald's introduction to Chapman, *Watkin's Last Expedition*, 11.

17. See above, chap. 2, n. 23.

18. In this descriptive poem, "Call of the North," the lines "I must go North again, for I / Have given my heart to Death," hark back to Kilmeny and to Isobel's own acceptance of death as the blessed state where her father and brothers can be reached and that she no longer fears.

5. HOUSEHOLDER IN GREENLAND

1. NLS, acc. 9713, no. 20, diary 1927.

2. NLS, acc. 9713, no. 21, diary 1928.

3. Royal Danish Ministry of Foreign Affairs, copy of letter from Grønlands Styrelse, no. 218, 19' Marts 1928, København, signed by Daugaard Jensen. Translated for me by Winnie Janzen.

4. NLS, acc. 9713, no. 21, diary 1928.

5. NLS, acc. 9713, no. 21, diary 1928, and Wm. Ketcham, "Science in the Shadow of the Pole," *Canadian Geographic Journal* 16 (March 1938): 138–41.

6. Kent, *Salamina*, 274.

7. Hutchison, *On Greenland's Closed Shore*, 132.

8. For a detailed geological description of the island of Umanak see Seward, *Summer in Greenland*, 88–89.

9. *Greenland*, 3:13–39.

10. Hutchison, *On Greenland's Closed Shore*, 136.

11. According to advertisements in the *Times* in 1907, in Britain the average annual wage for a general servant was £19 10s., and for a parlormaid, £26 8s., as quoted in Roberts, *Women's Work*, 32. Routh, *Occupation and Pay in Britain*, 95, notes that domestic servants averaged £49 a year in 1906 and £115 a year in 1924. In these terms, Dorthe's pay—14 s. a month, or £8 8s. a year—was a great bargain.

12. NLS, acc. 9713, no. 21, diary 1928.

13. *National Geographic* 167, no. 2 (1985): 191–201; Rosing, *The Sky Hangs Low*.

Jens Rosing was the director of the Greenland Museum in 1972 and took charge of the burial site. He is the son of Otto Rosing and was three years old at the time of Isobel Hutchison's year in Umanak.

14. NLS, acc. 8138, box 1, no. 9, diary August 1928–May 1929, Greenland.
15. Hutchison, *On Greenland's Closed Shore*, 178.
16. Hutchison, *On Greenland's Closed Shore*, 244.
17. Diebitsch-Peary, *My Arctic Journal*, 97, 99.
18. *Greenland*, 3:66.
19. Freuchen, *My Life in the Frozen North*, 251.
20. H. Jacobi, ed., "Laerinde; Umanak efteraret og vinteren 1899–1900; Lisbeth's grønlands dagbog," *Grønland*, no. 10 (1963): 384–92, translated for me by Winnie Janzen.
21. Hutchison, *On Greenland's Closed Shore*, 198.
22. Hutchison, *On Greenland's Closed Shore*, 154.
23. Rasmussen, *Greenland by the Polar Sea*; Mikkelsen, *Lost in the Arctic*.

6. UNKNOWN ISLAND

1. Hutchison, *On Greenland's Closed Shore*, 330.
2. For a discussion of reciprocal observations see Pratt, *Imperial Eyes*, especially her example of Mungo Park as antihero on page 81. The use of self-deprecating humor is discussed in LaFramboise, "Travellers in Skirts," 138–39.
3. Hutchison, *On Greenland's Closed Shore*, 347.
4. Hutchison, *On Greenland's Closed Shore*, 373.
5. Jackson, *British Whaling Trade*, chap. 3; *Greenland*, 3:19.
6. Kingsley, *Travels in West Africa*; Isabella Bird, *Six Months in the Sandwich Islands* (1875), as told in Barr, *Curious Life for a Lady*.
7. Nansen, *First Crossing of Greenland*, 419.
8. Ehrstrom, *Doctor's Wife in Greenland*, 79.
9. NLS, acc. 9713, no. 21, diary 1928.
10. Kent, *Salamina*, 260.

7. PRELUDE TO ADVENTURE

1. NLS, acc. 4775, letter to whom it may concern from Knud Rasmussen, Copenhagen, April 23, 1930.
2. NLS, acc. 9713, no. 23, diary 1931; acc. 8138, box 8, no. 1, Lists of lectures, 1928–67.
3. NLS, acc. 9713, no. 23, diary 1931.
4. NLS, acc. 9713, no. 23, diary 1931.
5. Archives Scott Polar Research Institute (hereafter SPRI), 069.41, IWH to Prof. Debenham, November 7, 1931.
6. NLS, acc. 4775, Frank Debenham, director Scott Polar Research Institute, to IWH, November 18, 1931.
7. SPRI, 069.41, IWH to Prof. Debenham, February 7, 1933.
8. *Polar Record* 1, no. 7 (1934): 74.

9. Royal Botanic Gardens, Kew, PRO 4/H/29, Mrs. [*sic*] Isobel W. Hutchison (hereafter Archives Kew), undated letter from IWH.

10. Archives Kew, PRO 4/H/29, John Gilmour, assistant director, to IWH, April 11, 1933.

11. Archives Kew, PRO 4/H/29, undated letter from IWH.

12. Archives Kew, PRO 4/H/29, IWH to Dr. Cotton, director, April 12, 1933.

13. Stefansson Collection, Dartmouth College, Hanover NH, Orcutt file, Stefansson to Orcutt, April 7, 1933; Hutchison file, IWH to Stefansson, April 17, 1933, and telegram IWH to Stefansson, April 19, 1933.

14. Stefansson Collection, Hutchison file, IWH, Carlowrie, to Stefansson, April 22, 1933.

15. Stefansson Collection, Stefansson to IWH, May 10, 1933.

16. National Archives of Canada (hereafter NAC), T-13348 file: 7834, IWH to secretary of the Department of the Interior, Vancouver, June 9, 1933.

17. NAC, T-13348 file: 7834, Memorandum from H. H. Rowatt, deputy minister, June 30, 1933.

18. NAC, T-13348 file: 7834, Memorandum from J. F. Doyle, Department of the Interior, August 24, 1933.

19. NAC, T-13348 file: 7834, J. F. Doyle to IWH, July 7, 1933.

20. NLS, acc. 9713, no. 25, diary 1933.

21. Coates, *Canada's Colonies*, chap. 4.

22. Hudson's Bay Company Archives (hereafter HBCA), Ships' histories: *Anyox*.

23. Hutchison, *North to the Rime-Ringed Sun*, 62.

24. Willoughby (Bill) Mason had been a trapper and prospector in Arctic and subarctic Canada and in Alaska for twenty years when he met and married a young woman who had come from New York to work at the Hudson Stuck Memorial Hospital in Fort Yukon. He is mentioned in Stefansson's book *The Friendly Arctic* and by many others writing about that area.

25. Hutchison, *North to the Rime-Ringed Sun*, 71.

26. NLS, acc. 9713, no. 20, diary 1927. The earlier trip was a flight from Copenhagen, Denmark, to Malmö, Sweden.

27. *Alaskans*, 139.

28. Hunt, *North of 53°*, 113.

29. Robertson, *Gentleman Adventurer*, 205.

30. Hrdlička, *Alaska Diary*, 81.

31. Lomen, *Fifty Years in Alaska*.

32. NLS, acc. 9713, no. 25, diary 1933.

33. Davis, *Uncle Sam's Attic*, 149.

34. LeBourdais, *Northward on the New Frontier*, 222–29.

35. Archives Kew, PRO 4/H/29, IWH to Dr. Cotton, Nome, Alaska, July 13, 1933.

36. Archives Kew, PRO 4/H/29, Dr. Cotton to IWH, September 16, 1933.

37. NLS, acc. 9713, no. 25, diary 1933.

38. Rasmussen, *Across Arctic America*, 348.

39. HBCA, B 378/a/2, Post records, Aklavik, 1933–34, July 27, 1933.

40. Hutchison, *North to the Rime-Ringed Sun*, 87.

41. *Saturday Review of Literature*, July 17, 1937.

8. INTO THE ICE

1. Irwin, *Alone across the Top of the World*, 11.
2. Irwin, *Alone across the Top of the World*, 12.
3. Irwin, *Alone across the Top of the World*, 26.
4. Conversation with Nessie McGowan, widow of the tenant farmer at Carlowrie, April 29, 1997.
5. Alling's story is told in many publications: Hoyle, "Women of Determination," 133; W. W. Bride in *Beaver*, September 1943, 15; *Dawson News*, October 6, 1928, 4; and John Stevenson in *Pioneer News* (Bank of British Columbia) 11, no. 5 (1988): 1.
6. Hutchison, *North to the Rime-Ringed Sun*, 101.
7. Bockstoce, *Whales, Ice and Men*, 344–45, 139.
8. Brower, *Fifty Years below Zero*, 72.
9. Bockstoce, *Whales, Ice and Men*, 27–37.
10. Hulley, *Alaska, Past and Present*, 202–7.
11. Stuck, *Winter Circuit of Our Arctic Coast*, 103.
12. Rasmussen, *Across Arctic America*, 329.
13. Mikkelsen, *Mirage in the Arctic*, 132.
14. Hutchison, *North to the Rime-Ringed Sun*, 104.
15. Hutchison, *North to the Rime-Ringed Sun*, 110.
16. HBCA, Ships' histories: *Baychimo*.
17. Robertson, *Gentleman Adventurer*, 181–82.
18. Information for the *Baychimo* story is from the Hudson's Bay Company Archives; the Richard Bonnycastle diaries in Robertson, *Gentleman Adventurer;* and the diary of James Hay, one of the skeleton crew remaining with the *Baychimo* on the Arctic coast, in *Beaver*, spring 1972, 44.
19. HBCA Post Records, Herschel Island, August 1933.
20. Hutchison, *North to the Rime-Ringed Sun*, 117.
21. Burberry is a waterproof cloth made by the company of that name. It was much favored in earlier polar travel, being used by all the members of Shackleton's team in 1914.
22. Archives of the Royal Scottish Geographical Society, Glasgow, box 7, Miscellaneous papers. This is the only personal letter found regarding any of her journeys, from IWH to her sister Hilda, undated and presumably posted in Barrow.
23. Brower, *Fifty Years below Zero*, 124, 182.
24. Hutchison, *North to the Rime-Ringed Sun*, 129.
25. This pottery bowl is on permanent display in the Arctic case at the Cambridge University Museum of Archaeology and Anthropology.

9. PRISONER ON A SANDSPIT

1. Details of the life of Gus Masik are from Isobel Hutchison's book *Arctic Nights' Entertainments*.
2. Hutchison, *North to the Rime-Ringed Sun*, 142.

3. Jenness, *Arctic Odyssey.*
4. Mikkelsen, *Mirage in the Arctic.*
5. Rasmussen, *Across Arctic America,* 304.
6. Stuck, *Winter Circuit of Our Arctic Coast,* 271.
7. Hutchison, *North to the Rime-Ringed Sun,* 149.
8. HBCA Post Records, Herschel Island, September 17, 1933.
9. Hutchison, *North to the Rime-Ringed Sun,* 151.
10. NLS, acc. 9713, no. 25, diary 1933.
11. Hutchison, *North to the Rime-Ringed Sun,* 152.
12. Hutchison, *North to the Rime-Ringed Sun,* 152–53.
13. NLS, acc. 9713, no. 25, diary 1933.
14. Hutchison, *North to the Rime-Ringed Sun,* 163.
15. Jenness, *Arctic Odyssey,* 16, 51–53, 115–22, etc.
16. HBCA Post Records, Herschel Island, October 3, 1933.
17. NAC, T-13348 file: 7834, Charles Rivett-Carnac, report to Royal Canadian Mounted Police (hereafter RCMP) headquarters in Ottawa, October 6, 1933, para. 5.
18. Ibid., para. 8.
19. Rivett-Carnac, *Pursuit in the Wilderness,* 271–72.
20. North, *Mad Trapper of Rat River,* and North, *Trackdown.*
21. HBCA Post Records, Herschel Island, October 19, 1933.
22. Hutchison, *North to the Rime-Ringed Sun,* 164.
23. *Times Literary Supplement,* May 23, 1935, 325.
24. Hutchison, *Arctic Nights' Entertainments,* 138.
25. Jenness, *Arctic Odyssey.*
26. For a full account of the reindeer drive and David Irwin's brief involvement see North, *Arctic Exodus.*
27. Irwin, *Alone across the Top of the World,* 101–3.
28. NLS, acc. 9713, no. 25, diary 1933.
29. Hutchison, *North to the Rime-Ringed Sun,* 167.
30. Hutchison, *North to the Rime-Ringed Sun,* 168.
31. HBCA Post Records, Herschel Island, October 23, 1933.
32. Hutchison, *North to the Rime-Ringed Sun,* 172.
33. Hutchison, *North to the Rime-Ringed Sun,* 182.
34. Hutchison, *North to the Rime-Ringed Sun,* 184.
35. NLS, acc. 9713, no. 25, diary 1933.
36. Only one newspaper story mentioned that Isobel spent a night with Masik in an igloo: "Miss Hutchison herself was storm bound on a long trek across waste lands. She and the dog musher had to hole up in an ice house. The temperature was 60 below. . . . Yet the inside of that hut . . . was snug and warm. Miss Hutchison didn't take all her clothes off. (I didn't ask her but I just know she didn't.) "Wife-Swapping Now Obsolete in the Arctic," *Toronto Daily Star,* February 24, 1934. Although the article is racy and dismissive, the writer, Gordon Sinclair, who made his reputation by asking intrusive personal questions, was obviously held at bay by Isobel's cool, ladylike manner.
37. Hutchison, *North to the Rime-Ringed Sun,* 185.

10. BY DOGSLED TO AKLAVIK

1. Nuligak, *I, Nuligak,* 9, 31–33.
2. Peake, *Bishop Who Ate His Boots,* 30–38.
3. Milloy, *Suffer the Little Children.*
4. Milloy, *Suffer the Little Children.*
5. HBCA B378/a/2, Post Records, Aklavik, November 29, 30, 1933.
6. Milloy, *Suffer the Little Children.*
7. North, *Arctic Exodus.*
8. Rivett-Carnac, *Pursuit in the Wilderness,* 258.
9. HBCA B378/a/2, Post Records, Aklavik, November 9, 1933, February 6, 1934.
10. Robertson, *Gentleman Adventurer,* 139.
11. NLS, acc. 9713, no. 25, diary 1933.
12. Rivett-Carnac, *Pursuit in the Wilderness,* 279.
13. A conversation with A. W. Laird of Vancouver, who served as a special constable in the RCMP in Saskatchewan in the late 1930s, revealed that Charles Rivett-Carnac was a very reasonable and approachable commanding officer.
14. North, *Lost Patrol.*
15. NLS, acc. 9713, no. 25, December 29, 30, diary 1933.
16. NAC, T-13348 file: 7834, memorandum to H. H. Rowatt, deputy minister, Department of the Interior, January 19, 1934, from acting chairman, Dominion Lands Board, J. Lorne Turner.
17. NAC, T-13348 file: 7834, IWH, Aklavik, January 3, 1934, to chairman, Dominion Lands Board.
18. NAC, T-13348 file: 7834, memorandum to H. H. Rowatt, January 19, 1934.
19. NAC, T-13348 file: 7834, IWH, Aklavik, January 3, 1934, to chairman, Dominion Lands Board.
20. NAC, T-13348 file: 7834, Charles Rivett-Carnac, Aklavik, November 28, 1933, official report to RCMP headquarters in Ottawa.
21. NLS, acc. 9713, no. 75, Account book, Alaska. Isobel Hutchison kept detailed records: the cost of her trip, from buying her clothing, camera, and so on, to the airplane ticket to leave Aklavik, was 478.17.6 in pounds sterling, or roughly $1,916.
22. NLS, acc. 9713, no. 25, December 8, 21, diary 1933.
23. NAC, T-13348 file: 7834, letter to IWH, January 25, 1934, from J. Lorne Turner, Dominion Lands Board, Ottawa.
24. North, *Arctic Exodus,* and communications with Charlene Porsild, granddaughter of Bob Porsild.
25. Hutchison, *North to the Rime-Ringed Sun,* 230.
26. Stefansson Collection, IWH, Hotel Taft, New York, to Stefansson, February 19, 1934.
27. Archives Kew, PRO 4/H/29, newspaper clippings.
28. Stefansson Collection, IWH, Carlowrie, to Stefansson, April 22, 1934.
29. Archives Kew, PRO 4/H/29, newspaper clippings.
30. Hoyle, "Women of Determination," 130.
31. Hutchison, *North to the Rime-Ringed Sun,* 237.

11. LURE OF DISTANT HORIZONS

1. Archives Kew, PRO 4/H/29, correspondence between IWH and Dr. Cotton and Sir Arthur Hill, May 3, May 4, and June 27, 1934.
2. Archives Kew, PRO 4/H/29, Minute sheet, July 25, 1934.
3. Archives Kew, PRO 4/H/29, IWH to Sir Arthur Hill, July 27, 1934.
4. Archives Kew, PRO 4/H/29, IWH to Sir Arthur Hill, July 27, 1934.
5. Archives Kew, PRO 4/H/29, Mr. Gilmour to IWH, July 4, 1934, unsigned.
6. Archives Kew, PRO 4/H/29, IWH to Sir Arthur Hill, July 27, 1934.
7. *Bulletin of Miscellaneous Information* (Royal Botanic Gardens, Kew), no. 9 (1934): 345–52; Archives Kew, PRO 4/H/29, IWH to Sir Arthur Hill, December 23, 26, 1934.
8. Archives Kew, PRO 4/H/29, IWH to Sir Arthur Hill, June 27, 1934.
9. NLS, acc. 4775, Stefansson to IWH, July 10, 1934.
10. NLS, acc. 4775, Stefansson to IWH, April 12, 1934.
11. NLS, acc. 4775, Louis Clarke to IWH, March 5, 1934.
12. NLS, acc. 4775, Capt. Cornwell to IWH, March 19, 1934.
13. Stefansson Collection, IWH at Carlowrie to Stefansson, July 24, 1934.
14. *Polar Record* 2, no. 15 (1937): 51.
15. *Times Literary Supplement,* November 22, 1934, 806; *Scottish Geographical Magazine* 51 (March 1935): 116–17.
16. Stefansson Collection, IWH to Stefansson, December 12, 1937.
17. Archives Kew, PRO 4/H/29, IWH to Sir Arthur Hill, December 23, 1934.
18. NLS, acc. 4775, Mrs. Greist to IWH, August 28, 1935; Ira Rank to IWH, March 10, 1935; H. M. Speechly, Manitoba Museum, February 25, 1935; Prof. Wm. Cooper, April 18, 1935.
19. NLS, acc. 5509, no. 2, Edith Tyrrell to IWH, December 18, 1935.
20. NLS, acc. 4775, Ira Rank to IWH, March 10, 1935; March 29, 1935.
21. NLS, acc. 4775, Charles Thornton to IWH, October 29, 1935.
22. Kew Archives, PRO 4/H/29, IWH to Sir Arthur Hill, December 4, 1935.
23. *Scottish Geographical Magazine* 51 (July 1935): 254.
24. Stefansson Collection, Stefansson to IWH, June 8, 1935.
25. NLS, acc. 4775, Mrs. Greist to IWH, August 28, 1935.
26. Stefansson Collection, IWH, on the *Disko,* to Stefansson, August 29, 1935; *Polar Record* 1, no. 11 (1935): 96.
27. NLS, acc. 9713, no. 27, diary 1935.
28. Kew Archives, PRO 4/H/29, IWH to Sir Arthur Hill, December 4, 1935.
29. NLS, acc. 4775, Stefansson to IWH, July 10, 1935.
30. Kew Archives, PRO 4/H/29, IWH to Sir Arthur Hill, December 4, 1935; Sir Arthur Hill to IWH, December 12, 1935, unsigned.
31. Kew Archives, PRO 4/H/29, IWH to Sir Arthur Hill, December 16, 1935, April 30, 1936.
32. British Museum Archives, DF 400/51, IWH to Mr. Ramsbottom, April 27, 1936; Mr. Ramsbottom to IWH, April 29, 1936; IWH to Mr. Ramsbottom, May 4, 1936.
33. Laughlin, *Aleuts,* 10–20.

34. Hubbard, *Cradle of the Storms*, 27; Hunt, *Arctic Passage*, chap. 8.
35. Walker, *Working on the Edge*.
36. Egbert Walker, "Plants of the Aleutian Islands," in *The Aleutian Islands: Their People and Natural History*, by Henry B. Collins (Washington DC: Smithsonian Institution, 1945).
37. Hunt, *Arctic Passage*, 264–66; and McCracken, *God's Frozen Children*, 49.
38. Hubbard, *Cradle of the Storms*, 210–32.
39. Hutchison, *Aleutian Islands* (wartime edition of *Stepping Stones from Alaska to Asia*), 60.
40. Hubbard, *Cradle of the Storms*, 143–44, 182–83.
41. Hubbard, *Cradle of the Storms*, 51–52, 58–60.

12. TO THE EDGE OF THE WESTERN WORLD

1. John, *Libby*.
2. Hutchison, *Aleutian Islands*, 101.
3. Hulley, *Alaska, Past and Present*, 206.
4. See Jochelson, *History, Ethnology and Anthropology of the Aleut*; McCracken, *God's Frozen Children*.
5. Hutchison, *Aleutian Islands*, 114.
6. Collins, *Aleutian Islands*, 1–30; Oliver, *Journal of an Aleutian Year*, 45–48, 57–59.
7. The murder is described in detail in Wheaton, *Prekaska's Wife*, 216–33. Helen Wheaton hated every moment of her time on Atka and poured her venom into a book about the experience, disparaging the people of the village as dirty, drunken, and degraded. Unfortunately her book, published in 1945, was read on Atka, and people were hurt and offended by her opinions, as the Olivers learned when they arrived the following year to take charge of the school and local administration. Ethel Oliver describes the reaction of the Atkans in *Journal of an Aleutian Year*, 67.
8. Oliver, *Journal of an Aleutian Year*, 37.
9. Wheaton, *Prekaska's Wife*, 171–91.
10. Wheaton, *Prekaska's Wife*, 176.
11. Oliver, *Journal of an Aleutian Year*, xvii, 84.
12. Nutchuk and Hatch, *Son of the Smoky Sea*, 38, 44–57.
13. *Geographical Journal* 90 (December 1937): 541–46.
14. Oliver, *Journal of an Aleutian Year*, 220–54.
15. Oliver, *Journal of an Aleutian Year*, xviii.
16. Hutchison, *Aleutian Islands*, 167–68.
17. British Museum Archives, DF 400/51, IWH to Ramsbottom, September 14, 1936.
18. From the Hutchison collection, Alaska Film Archives, University of Alaska, Fairbanks.
19. Hutchison, *Aleutian Islands*, 174.
20. Hunt, *Arctic Passage*, 261–81.
21. Hutchison, *Aleutian Islands*, 174.

22. Hutchison, *Aleutian Islands,* 176.
23. Hultén, *Flora of the Aleutian Islands,* 9–19, 370–75.
24. *Times Literary Supplement,* December 4, 1937, 923.
25. *Nature* 139 (February 20, 1937): 327.
26. *Nature* 141 (March 12, 1938): 465.

13. AROUND THE WORLD AND HOME AGAIN

1. NLS, acc. 9713, no. 28, diary 1936; Archives Kew, PRO 4/H/29, IWH to Sir
    Arthur Hill, March 6, 1937; British Museum Archives, DF 400/51, IWH to Mr.
    Ramsbottom, December 26, 1936; NLS, acc. 4775, Alan May to IWH, April 5,
    1937; Ira Rank to IWH, May 5, 1937.
2. *Times* (London) (all 1937): February 15, 13e; February 16, 17e; February 27,
    13e.
3. "The Discovery of a New Reef Near Attu Island," *Geographical Journal* 90
    (December 1937): 541–46.
4. Archive, Royal Geographical Society (hereafter RGS), corr.bl. 1931–40.
5. Barr, *Curious Life for a Lady,* 333.
6. *Times* (London) (all 1893): May 19, 7c; Lord Curzon's letter, May 31, 11d; and
    further correspondence, June 1, 4e; June 3, 6d; June 5, 10b; June 10, 12e; July
    1, 6f; July 4, 12a; August 26, 8f.
7. Birkett, *Spinsters Abroad,* 219, quoting from the *Scottish Geographical Magazine*
    1 (1885): 18.
8. Birkett, *Spinsters Abroad,* 215, 218.
9. NLS, acc. 4775, RGS to IWH, December 16, 1937, and January 21, 1938.
10. Color film, introduced in 1935 for thirty-five-millimeter cameras, did not come
    into general use until 1942. It is most likely that the film Isobel used in 1936
    had shades of gray, brown, and green added in the laboratory.
11. NLS, acc. 4775, Stefansson to IWH, May 10 and November 23, 1937.
12. NLS, acc. 9713, no. 69, Letters and material relating to Gus Masik, and no. 29,
    diary 1937.
13. NLS, acc. 8138, box 5, no. 1, box 8, no. 4; acc. 9713, no. 30, diary 1938.
14. *Times* (London), October 15, 1938, 13f and 14a.
15. NLS, acc. 9713, no. 31, diary 1939.
16. NLS, acc. 9713, no. 31, diary 1939; conversation with Nessie McGowan, April
    29, 1997.
17. Fax from BBC Written Archives Centre, Reading, England, March 23, 1998;
    NLS, acc. 9713, no. 34, diary 1942. The Royal Scottish Geographical Society
    Archives (hereafter RSGS) have scripts of all the radio broadcasts.
18. Letter from Mary Anne McMillen, director, Records Library, National Geo-
    graphic Society, Washington DC, March 15, 1997.
19. Capt. Kielhorn to IWH, May 28, 1941, quoted in Hutchison, *Aleutian Islands;*
    RSGS Archives, Correspondence file, Capt. Kielhorn to IWH, July 8, 1942,
    enclosed in a letter from Rear Admiral Chalker.
20. Stefansson Collection, IWH to Reg Orcutt, April 9, 1951.
21. NLS, acc. 9713, no. 38, diary 1946.

22. Morgan, *People's Peace,* 68–69, and Morgan, *Labour in Power,* 347–65.

23. I was living in Britain from 1950 to 1952, and for a trip to continental Europe the currency allowance regardless of the length of the visit was ten pounds sterling per person.

24. NLS, acc. 9713, no. 39, diary 1947.

25. *Weekly Scotsman* (Edinburgh), June 23, 1949, interview by Jean Speedy.

26. NLS, acc. 9713, no. 41, diary 1949; acc. 4775, Letters of congratulation to IWH from Ejnar Mikkelsen; from Dorthe in Greenland; from Capt. Kielhorn; from Alan May, now at Harvard; from A. E. Porsild; and many others.

27. Robinson, *Wayward Women,* 135.

28. Stefansson Collection, IWH to Stefansson, June 27, 1947.

29. NLS, acc. 8138, box 9, nos. 1, 2, "Literary Venture" notebooks.

30. NLS, acc. 4775.

31. NLS, acc. 9713, no. 42, diary 1950; acc. 5509, no. 3, Amy von Hollitcher to IWH, 1952, Asolo, describing Stark as "inclined to be superior about other peoples works but I could see that she was impressed by the watercolours [in *North to the Rime-Ringed Sun*]."

32. I had a similar experience when hiking across the open moors in the Highlands. A rain squall struck, and there was nothing larger than a bush for shelter. We were close to a hotel marked on the map, but when we approached it was obviously no longer a hostelry, and the farmer did not welcome visitors.

33. RSGS Archives, Articles file.

34. Tiltman, *Women in Modern Adventure.*

35. This was mentioned in the *Times Literary Supplement* review of *On Greenland's Closed Shores.*

36. RGS Archive, March 9, 1937.

37. NLS, acc. 8138, box 8, no. 1, Lists of lectures, 1928–67.

38. *Polar Record* 1, no. 11 (1935): 39–40.

39. NLS, acc. 9713, box 68, IWH from St. Andrew's Presbyterian Church, Jerusalem, 1961; NLS, acc. 9713, box 76, Lists of donations to charities, 1959–71.

40. NLS, acc. 9713, box 68, IWH from Ejnar Mikkelsen, January 13 and February 6, 1936.

41. RSGS Archives, Travel notes file.

42. NLS, acc. 9713, no. 46, diary 1954.

43. Personal letter to the author from James Baird, Edinburgh, January 21, 1998.

44. NLS, acc. 9713, nos. 42–50, diaries, 1950–58.

14. BLESSINGS OF FRIENDSHIP, CURSE OF OLD AGE

1. NLS, acc. 5509, no. 3, Palsson to IWH, March 27, 1935.

2. NLS, acc. 5509, no. 3, John Glynn, officer on the *Chelan,* 1937.

3. NLS, acc. 9713, no. 68, Maynard Williams to IWH, 1955.

4. Conversation with Nessie McGowan, April 30, 1997.

5. Jean Speedy, "A Woman Explorer Whom St. Andrews Will Honour," *Weekly Scotsman,* June 23, 1949.

6. Matlin, *Psychology of Women,* 280.

7. Morrow, *Names of Things*, 95.
8. Recent books dealing with this topic: Simon, *Never Married Women*; Chambers-Schiller, *Liberty, a Better Husband*; Vicinus, *Independent Women*; and Clements, *Improvised Woman*.
9. Dyhouse, *Feminism and the Family in England*, 159.
10. Hutchison, *Original Companions*, 80.
11. NLS, acc. 9713, no. 68, Horoscope.
12. NLS, acc. 8138, box 5, no. 9, Miscellaneous articles, 1922–44, Notes of a woman explorer requested by the *Glasgow Herald* in 1936: "It is in our stars—my horoscope said I was fated to visit the Arctic."
13. NLS, acc. 9713, no. 68, Dorothy Allard to IWH, January 1982.
14. NLS, acc. 9713, no. 68, Marian Harvey to IWH, July 1934.
15. NLS, acc. 9713, no. 69, the file of letters from Gus Masik; and Beulah Hermann, Bellevue, Washington, to IWH, May 17, 1976.
16. Conversation with Nessie McGowan, April 30, 1997.
17. Conversation with Rev. John Murrie, October 4, 1998.
18. Conversation with Mr. and Mrs. James Gammell, October 15, 1998.
19. Conversation with Katrina Burnett, niece of Medina Lewis, October 21, 1998.
20. NLS, acc. 8138, box 9, no. 1, "Literary Venture" notebooks, 1910–44.
21. Communication from Nicholas Martland, enquiries and research support librarian, Royal Botanic Gardens, Kew, April 7, 1997.
22. When looking at mounted specimens of primula in the herbarium of the Royal Botanic Garden, Edinburgh, I noticed that Isobel Hutchison's flowers were filed next to some collected by Dr. John Richardson on the Franklin land expedition in 1822.
23. Discussion with Briony Crozier, curator of North American artifacts, Royal Scottish Museum, Edinburgh, October 2, 1998.
24. NLS, acc. 9713, no. 50, diary 1958.

APPENDIX 1. THE LITERATURE OF TRAVEL AND ADVENTURE

1. All references to published material are included in the bibliography.
2. The list of secondary sources on this subject could run to many pages. In addition to the items mentioned above, see the following examples of this analytical material: Judith Adler, "Travel as Performed Art," *American Journal of Sociology* 94 (1989): 1366–91; Ali Behdad, *Orientalism in the Age of Colonial Dissolution* (Durham NC: Duke University Press, 1994); Alison Blunt, *Travel, Gender, and Imperialism: Mary Kingsley and West Africa* (New York: Guilford, 1994); Shirley Foster, *Across New Worlds: Nineteenth Century Women Travellers and Their Writings* (Hemel Hempstead, Hertfordshire: Harvester Wheatsheaf, 1990); Maria Frawley, *A Wider Range: Travel Writing by Women in Victorian England* (Rutherford NJ: Fairleigh Dickinson University Press, 1994); Les Harding, *The Journeys of Remarkable Women: Their Travels on the Canadian Frontier* (Waterloo ON: Escart, 1994); Karen Lawrence, *Penelope Voyages: Women and Travel in the British Literary Tradition* (Ithaca: Cornell University Press, 1994); Dennis Porter, *Haunted Journeys: Desire and*

Transgression in European Travel Writing (Princeton: Princeton University Press, 1991); and Edward Said, *Orientalism* (New York: Random House Vintage, 1994).

3. Although their books do not fit the category of travel literature, Black's *My Seventy Years* and Berton's *I Married the Klondike* are part of the Yukon canon.

INDEX

1 When cross-references have very few page numbers, we normally repeat the page numbers in both entries.

2 Because the entry that is referenced contains only one additional page number, I simply added it here.

Bibliography

PUBLICATIONS OF ISOBEL WYLIE HUTCHISON

Books

POETRY

Lyrics from West Lothian. Privately published, 1916.
How Joy Was Found: A Fantasy in Verse in Five Acts. London: Blackie; New York: Frederick A. Stokes, 1917.
The Calling of Bride. Stirling, Scotland: E. Mackay, 1926.
The Song of Bride. London: De La More, 1927.
The Northern Gate. London: De La More, 1927.
Lyrics from Greenland. London: Blackie, 1935.

PROSE

Original Companions. London: Bodley Head, 1923.
The Eagle's Gift: Alaska Eskimo Tales. Translation of *Festen Gave,* by Knud Rasmussen. Garden City, N.Y.: Doubleday, Doran, 1932.
Flowers and Farming in Greenland. Edinburgh: T. A. Constable, 1930. Reprinted from *Scottish Geographical Magazine,* vol. 66 (July 1930).
On Greenland's Closed Shore: The Fairyland of the Arctic. Edinburgh: William Blackwood, 1930.
North to the Rime-Ringed Sun: Being the Record of an Alaska-Canadian Journey Made in 1933–34. London: Blackie, 1934, 1935; New York: Hillman-Curl, 1937.
[With August Masik]. *Arctic Nights' Entertainments: Being the Narrative of an Alaskan Estonian Digger, August Masik, as Told to Isobel Wylie Hutchison.* Glasgow: Blackie, 1935.
Stepping Stones from Alaska to Asia. London: Blackie, 1937. Reissued as *The Aleutian Islands: America's Back Door.* London: Blackie, 1942.

Articles

NATIONAL GEOGRAPHIC

"Walking Tour across Iceland," April 1928
"Riddle of the Aleutians," December 1942

"Scotland in Wartime," June 1943
"Wales in Wartime," June 1944
"Bonnie Scotland, Post-war Style," May 1946
"2000 Miles through Europe's Oldest Kingdom," February 1949
"A Stroll to London," August 1950
"A Stroll to Venice," September 1951
"Shetland and Orkney, Britain's Far North," October 1953
"From Barra to Butt in the Hebrides," October 1954
"A Stroll to John o'Groat's," July 1956
"Poets' Voices Linger in Scottish Shrines," October 1957

OTHER MAGAZINES, NEWSPAPERS, AND JOURNALS THAT PUBLISHED HER POEMS AND
ARTICLES

*American Red Cross, American-Scandinavian Review, An Gaidhead, Blackwood's,
Bookman, Chambers's Journal, Contemporary Review, Cornhill, Daily Mail,
Geographic Journal, Glasgow Herald, Hors Concours, Jongleur, Manchester
Guardian, Nature, Observer, Poetry Review, Polar Record, Scottish Field,
Scottish Geographical Magazine, Scotsman, Scribner's, Spectator, Sphere,
Times of London, Westminster Gazette,* and many others.

ARCHIVES

Alaska Film Archives, University of Alaska, Fairbanks
British Broadcasting Corporation, Written Archives Centre, Reading
British Museum Archives, London
Cambridge University, Museum of Archaeology and Anthropology
Explorers Club, New York
Hudson's Bay Company Archives, Winnipeg, Manitoba
National Archives of Canada, Ottawa, Ontario
National Library of Scotland, Edinburgh
National Geographic Society, Washington DC
Natural History Museum of the British Museum, London
Royal Botanic Gardens, Kew (cited as Kew Archives)
Royal Botanic Garden, Edinburgh
Royal Geographical Society, London
Royal Scottish Geographical Society, Glasgow
Royal Scottish Museum, Edinburgh
Scott Polar Research Institute, Cambridge University
Stefansson Collection, Dartmouth College, Hanover, New Hampshire

SELECTED BIBLIOGRAPHY

The Alaskans. Text by Keith Wheeler, ed. Thomas H. Flaherty Jr. Alexandria VA:
Time-Life, 1977.

Auden, W. H., and Louis MacNiece. *Letters from Iceland.* 1937. London: Faber, 1967.

Backhouse, Frances. *Women of the Klondike.* Vancouver: Whitecap, 1995.

Banks, Theodore. *Wild Geese.* New Haven: Yale University Press, 1921.

Barr, Pat. *A Curious Life for a Lady: The Story of Isabella Bird, Traveller Extraordinary.* London: Penguin, 1985.

Bell, Gertrude. *The Desert and the Sown.* London: Heinemann, 1907.

———. *The Letters of Gertrude Bell:* Selected and edited by Lady Bell. London: E. Benn, 1927.

Bentley, Nicholas. *Edwardian Album.* New York: Viking Press, 1974.

Birkett, Dea. *Spinsters Abroad: Victorian Lady Explorers.* Oxford: Basil Blackwell, 1989.

Bloom, Lisa. *Gender on Ice: American Ideologies of Polar Expeditions.* Minneapolis: University of Minnesota Press, 1993.

Bockstoce, John R. *Whales, Ice, and Men.* Seattle: University of Washington Press, 1986.

Boswell, James. *Journal of a Tour to the Hebrides with Samuel Johnson.* London: Oxford University Press, 1924.

Brower, Charles D. *Fifty Years below Zero: A Lifetime Adventure in the Far North.* New York: Dodd, Mead, 1942.

Cameron, Agnes Deans. *The New North.* New York: Appleton, 1909.

Campbell, Joseph. *The Hero with a Thousand Faces.* Princeton: Princeton University Press, 1968.

Chambers-Schiller, Lee Virginia. *Liberty, a Better Husband.* New Haven: Yale University Press, 1984.

Chapman, Spencer. *Watkin's Last Expedition.* London: Vanguard, 1953.

Charcot, Dr. Jean. *The Voyage of the "Why Not?" in the Antarctic.* London: Hurst, 1978.

Clements, Marcelle. *The Improvised Woman.* New York: Norton, 1998.

Coates, Kenneth. *Canada's Colonies: A History of the Yukon and Northwest Territories.* Toronto: James Lorimer, 1985.

Collins, Henry B., Jr. *The Aleutian Islands: Their People and Natural History.* Washington DC: Smithsonian Institution, 1945.

Craig, Lulu Alice. *Glimpses of Sunshine and Shade in the Far North.* Cincinnati: Editor, 1900.

Crow, Duncan. *The Edwardian Woman.* London: George Allen and Unwin, 1978.

David-Neel, Alexandra. *My Journey to Lhasa.* London: Heinemann, 1927.

Davidson, Robyn, *Desert Places.* London: Penguin, 1996.

———. *Tracks.* London: Jonathan Cape, 1980.

Davis, Mary Lee. *Uncle Sam's Attic.* Boston: W. A. Wilde, 1930.

Diebitsch-Peary, Josephine. *My Arctic Journal: A Year among Ice-Fields and Eskimos.* London: Longmans Green, 1893.

Dyhouse, Carol. *Feminism and the Family in England, 1880–1939.* Oxford: Basil Blackwell, 1989.

Ehrstrom, Inga. *Doctor's Wife in Greenland.* Trans. F. H. Lyon. London: Allen and Unwin, 1955.

Freuchen, Peter. *My Life in the Frozen North.* New York: Farrar and Rinehart, 1935.

Fussell, Paul. *Abroad: British Literary Travelling between the Wars.* New York: Oxford University Press, 1980.

———, ed. *The Norton Book of Travel.* New York: W. W. Norton, 1987.

Geniesse, Jane F. *Passionate Nomad: The Life of Freya Stark.* New York: Random House, 1999.

Gilmore, Leigh. *Autobiographics: A Feminist Theory of Women's Self-Representation.* Ithaca: Cornell University Press, 1994.

Graham, Stephen. *Tramping with a Poet in the Rockies.* New York: Appleton, 1922.

Greenland. Ed. Kristjan Bure. Copenhagen: Royal Danish Ministry for Foreign Affairs, 1961.

Greenland. Vol. 3. Ed. M. Vahl. Commission for the Direction of the Geological and Geographical Investigations of Greenland. London: Oxford University Press, 1929.

Grenfell, Wilfred T. *Labrador Looks at the Orient: Notes of Travel in the Near and Far East.* Boston: Houghton Mifflin, 1928.

Groves, David. *James Hogg: The Growth of a Writer.* Edinburgh: Scottish Academic Press, 1988.

Hargrave, Letitia. *The Letters of Letitia Hargrave, 1813–54.* Toronto: Champlain Society, 1947.

Hasluck, Lady Alexandra. *Portrait with Background: A Life of Georgiana Molloy.* Melbourne: Oxford University Press, 1966.

Hitchcock, Mary. *Two Women in the Klondike.* New York: G. P. Putnam's Sons, 1899.

Hodgins, Bruce, and Margaret Hobbs, eds. *Nastawgan: The Canadian North by Canoe and Snowshoe.* Toronto: Betelgeuse, 1985.

Hogg, James. *Selected Poems.* Ed. Douglas Mack. Oxford: Oxford University Press, 1970.

Hoyle, Gwyneth. "Women of Determination," in *Nastawgan: The Canadian North by Canoe and Snowshoe,* edited by Bruce Hodgins and Margaret Hobbs. Toronto: Betelgeuse, 1985.

Hrdlička, Aleš. *Alaska Diary.* Lancaster PA: Jacques Cattell, 1943.

Hubbard, Bernard, S.J. *Cradle of the Storms.* New York: Dodd, Mead, 1935.

Hubbard, Mina. *A Woman's Way through Unknown Labrador.* London: Murray, 1908.

Hulley, Clarence C. *Alaska, Past and Present.* 1953. Portland OR: Binfords and Mort, 1970.

Hultén, Eric. *The Flora of the Aleutian Islands and Westernmost Alaska Peninsula.* 2d ed. Weinheim NY: J. Cramer, 1960.

Hunt, William R. *Arctic Passage: The Turbulent History of the Land and People of the Bering Sea, 1697–1975.* New York: Charles Scribner's Sons, 1975.

———. *North of 53.* New York: Macmillan, 1974.

———. *Stef: A Biography of Vilhjalmur Stefansson.* Vancouver UBC Press, 1986.

Iceland. Cambridge: Naval Intelligence, 1942.

Irvine, Lucy. *Castaway.* London: Gollancz, 1983.

Irwin, David. *Alone across the Top of the World.* As told to Jack O'Brien. Chicago: John C. Winston, 1935.

Jackson, Gordon. *The British Whaling Trade.* London: Adam and Charles Black, 1978.

Jacobs, Jane, ed. *A Schoolteacher in Old Alaska: The Story of Hannah Breece.* Toronto: Random House, 1995.

Jason, Victoria. *Kabloona in the Yellow Kayak.* Winnipeg: Turnstone Press, 1995.

Jenness, Diamond. *Arctic Odyssey.* Ed. Stuart Jenness. Hull PQ: Canadian Museum of Civilization, 1991.

———. *Dawn in Arctic America.* Minneapolis: University of Minnesota Press, 1957.

Jochelson, Waldemar. *History, Ethnology and Anthropology of the Aleut.* Washington DC: Carnegie Institute, 1933.

John, Betty. *Libby: The Sketches, Letters and Journal of Libby Beaman, Recorded in the Pribilof Islands, 1879–1880.* Tulsa OK: Council Oak Books, 1987.

Keith, Marian. *Under the Grey Olives.* Toronto: McClelland and Stewart, 1927.

Kelcey, Barbara E. "Jingo Belles, Jingo Belles, Dashing through the Snow: White Women and Empire on Canada's Northern Frontier." Ph.D. diss., University of Manitoba, Winnipeg, 1994.

Kempe, Margery. *The Book of Margery Kempe, 1436.* Modern version by W. Butler-Bowdon. London: Oxford University Press, 1954.

Kent, Rockwell. *Salamina.* New York: Harcourt Brace, 1935.

Kidson, Peter. *Iceland in a Nutshell.* Reykjavik: Ferdahandbaekur, 1966.

Kingsley, Mary. *Travels in West Africa.* Abridged by Elspeth Huxley. London: Dent, 1987.

LaFramboise, Lisa. "Travellers in Skirts: Women in English-Language Travel Writing in Canada, 1820–1926." Ph.D. diss., University of Alberta, Edmonton, 1998.

Laughlin, William S. *Aleuts: Survivors of the Bering Land Bridge.* New York: Holt, Rinehart, Winston, 1980.

LeBourdais, D. M. *Northward on the New Frontier.* Ottawa: Graphic, 1931.

Lewis, Eiluned. *Dew on the Grass.* Woodbridge, Suffolk: Boydell, 1984.

Lewis, Peter. *Eveline: An Account of Mrs. Lewis, M.A., J.P.* Llandysul, Dyfed: Gomer, 1986.

Livingstone, William P. *Mary Slessor of Calabar.* London: Hodder and Stoughton, 1916.

Lomen, Carl. *Fifty Years in Alaska.* New York: David McKay, 1954.

MacLaren, Ian S., and Lisa LaFramboise, eds. *The Ladies, the Gwich'in and the Rat.* Edmonton: University of Alberta Press, 1998.

MacPherson, John. *Tales of Barra Told by the Coddy, John MacPherson.* Ed. J. L. Campbell. Edinburgh: Johnston and Bacon, 1960.

Manning, Ella. *An Igloo for the Night.* Toronto: University of Toronto Press, 1946.

Marsden, Kate. *On Sledge and Horseback to Outcast Siberian Lepers.* London: Record Press, 1893; London: Century Hutchinson, 1986.

Matlin, Margaret. *The Psychology of Women.* New York: Holt, Rinehart and Winston, 1987.

McCracken, Harold. *God's Frozen Children.* New York: Doubleday Doran, 1930.

Mee, Margaret. *In Search of Flowers of the Amazon Forest.* Ed. Tony Morrison. Woodbridge, Suffolk: Nonesuch Expeditions, 1988.

Merrick, Elliott. *Northern Nurse.* New York: Charles Scribner's Sons, 1943.

———. *True North.* Lincoln: University of Nebraska Press 1989.

Mikkelsen, Ejnar. *Lost in the Arctic.* London: Heinemann, 1921.

———. *Mirage in the Arctic.* London: Hart-Davis, 1955.

Miller, Jane. *Rebel Women: Feminism, Modernism and the Edwardian Novel.* London: Virago, 1994.

Milloy, John. *A National Crime: The Canadian Government and the Residential School System, 1879–1986.* Winnipeg: University of Manitoba Press, 1999.

———. *Suffer the Little Children.* On CD-Rom, Royal Commission on Aboriginal Peoples, *For Seven Generations: An Information Legacy of the Royal Commission on Aboriginal Peoples.* Ottawa: Libraxus, 1997.

Mills, Sara. *Discourses of Difference.* London: Routledge, 1991.

Morgan, Kenneth. *Labour in Power, 1945–51.* Oxford: Oxford University Press, 1984.

———. *The People's Peace: British History, 1945–89.* Oxford: Oxford University Press, 1990.

Morris, Mary, ed. *The Virago Book of Women Travellers.* London: Virago, 1994.

Morris, William. *Icelandic Journals.* London: Mare's Nest, 1996.

Morrow, Susan Brind. *The Names of Things.* New York: Riverhead Books, 1997.

Moss, John, ed. *Echoing Silence: Essays on Arctic Narrative.* Ottawa: University of Ottawa Press, 1995.

Murie, Margaret E. *Two in the Far North.* Anchorage: Alaska Northwest, 1978.

Murphy, Dervla. *Full Tilt.* London: Arrow, 1987.

Nansen, Fridtjof. *The First Crossing of Greenland.* London: Longmans Green, 1910.

New Worlds: Discovering and Constructing the Unknown in Anglophone Literature. Ed. M. Kuester, G. Christ, and R. Beck. Munich: Ernst Vogel, 2000.

Noice, Harold. *With Stefansson in the Arctic.* New York: Dodd Mead, 1923.

North, Dick. *Arctic Exodus: The Last Great Trail Drive.* Toronto: Macmillan, 1991.

———. *The Lost Patrol.* Anchorage: Alaska Northwest, 1978.

———. *The Mad Trapper of Rat River.* Toronto: Macmillan, 1972.

———. *Trackdown: The Search for the Mad Trapper.* Toronto: Macmillan, 1989.

North, Marianne. *Recollections of a Happy Life.* Ed. Mrs. J. A. Symonds. London: Macmillan, 1892.

———. *A Vision of Eden.* New York: Holt, Rinehart and Winston, 1980.

Nuligak. *I, Nuligak.* Trans. and ed. Maurice Metayer. Toronto: Peter Martin, 1966.

Nutchuk (Simeon Oliver), with Alden Hatch. *Son of the Smoky Sea.* New York: Julian Messner, 1941.

Oliver, Ethel Ross. *Journal of an Aleutian Year.* Seattle: University of Washington Press, 1988.

Oliver, Simeon. *Son of the Smoky Sea.* New York: Julian Messner, 1941.

Peake, Frank. *The Bishop Who Ate His Boots.* Toronto: Anglican Church of Canada, 1966.

Peary, Robert. *The North Pole.* New York: Greenwood, 1910, 1968.

Pfeiffer, Ida. *The Last Travels of Ida Pfeiffer.* Trans. H. W. Dulcken. New York: Harper, 1861.

Pratt, Mary Louise. *Imperial Eyes: Travel Writing and Transculturation.* London: Routledge, 1992.

Raby, Peter. *Bright Paradise: Victorian Scientific Travellers.* London: Random House, 1997.

Rasmussen, Knud. *Across Arctic America.* 1927. New York: Greenwood, 1969.

———. *Greenland by the Polar Sea.* London: Heinemann, 1921.

Rivett-Carnac, Charles. *Pursuit in the Wilderness.* Boston: Little, Brown, 1965.

Roberts, Elizabeth. *Women's Work, 1840–1940.* London: Macmillan, 1988.

Robertson, Heather. *A Gentleman Adventurer: The Arctic Diaries of Richard Bonnycastle, 1928–31.* Toronto: Lester and Orpen Dennys, 1984.

Robinson, Jane. *Unsuitable for Ladies: An Anthology of Women Travellers.* Oxford: Oxford University Press, 1994.

———. *Wayward Women: A Guide to Women Travellers.* Oxford: Oxford University Press, 1990.

Rosing, Jens. *The Sky Hangs Low.* Trans. Naomi Jackson Groves. Moonbeam ON: Penumbra Press, 1986.

Routh, Guy. *Occupation and Pay in Great Britain, 1906–60.* Cambridge: Cambridge University Press, 1965.

Russell, Mary. *The Blessings of a Good Thick Skirt*. London: Collins, 1988.

Sanecki, Kay. *A Short History of Studley College*. Studley College, 1990.

Scoresby, William. *An Account of the Arctic Regions with a History and Description of the Northern Whale-Fishery*. London: Archibald Constable, 1820. Reprinted Newton Abbot, Devon: David and Charles (Holdings), 1969.

Selby, Bettina. *Riding the Desert Trail*. London: Abacus, 1988.

———. *Riding the Mountains Down*. London: Victor Gollancz, 1984.

———. *Pilgrim's Road*. London: Abacus, 1994.

Seward, A. C. *A Summer in Greenland*. Cambridge: Cambridge University Press, 1922.

Simon, Barbara Levy. *Never Married Women*. Philadelphia: Temple University Press, 1987.

Skelton, Robin. *The Poet's Calling*. London: Heinemann, 1975.

Smith, Nelson C. *James Hogg*. Boston: Twayne, 1980.

Stark, Freya. *Dust in the Lion's Paw*. London: J. Murray, 1961.

———. *Traveller's Prelude: Autobiography, 1893–1927*. London: Century, 1983.

———. *The Valley of the Assassins*. London: J. Murray, 1934.

Stuck, Hudson. *A Winter Circuit of Our Arctic Coast*. New York: Charles Scribner's Sons, 1920.

Sutherland, Halliday. *Hebridean Journey*. London: Geoffrey Bles, 1939.

Tagore, Rabindranath. *Collected Poems and Plays*. London: Macmillan, 1962.

Taylor, Elizabeth. *The Far Islands and Other Cold Places*. St. Paul MN: Pogo, 1997.

Thayer, Helen. *Polar Dream*. New York: Doubleday, 1993.

Tiltman, Marjorie. *Women in Modern Adventure*. London: Harrap, 1935.

Traill, Catherine Parr. *Canadian Wild Flowers*. Montreal: J. Lovell, 1868.

Trollope, Joanna. *Britannia's Daughters: Women of the British Empire*. London: Hutchinson, 1983.

Turner, Louis. *The Golden Hordes: International Tourism*. London: Constable, 1975.

Vicinus, Martha. *Independent Women: Work and Community for Single Women, 1850–1920*. Chicago: University of Chicago Press, 1985.

Vyvyan, Clara. *Arctic Adventure*. London: Peter Owen, 1961.

Wallace, Dillon. *The Long Labrador Trail*. New York: Outing, 1907.

———. *The Lure of the Labrador Wild*. New York: Fleming Revell, 1905.

Wallach, Janet. *Desert Queen: The Extraordinary Life of Gertrude Bell, Adventurer, Adviser to Kings, Ally to Lawrence of Arabia*. New York: Anchor Books, 1991.

Walker, Spike, *Working on the Edge: Surviving in the World's Most Dangerous Profession, King Crab Fishing on Alaska's High Seas*. New York: St. Martin's Press, 1991.

Warwick, Frances, Countess of. *Life's Ebb and Flow*. London: Hutchinson, 1929.

Webb, Melody. *The Last Frontier.* Albuquerque: University of New Mexico Press, 1985.

Wheaton, Helen. *Prekaska's Wife: A Year in the Aleutians.* New York: Dodd Mead, 1945.

Whitelegg, Elizabeth, ed. *The Changing Experience of Women.* Oxford: Basil Blackwell, 1982.

Women's History: Britain, 1850–1945. London: University College London Press, 1995.

Young, George. *Tourism: Blessing or Blight?* London: Penguin, 1973.

Index

In the Women in the West series

When Montana and I Were Young:
A Frontier Childhood
By Margaret Bell
Edited by Mary Clearman Blew

Martha Maxwell, Rocky Mountain Naturalist
By Maxine Benson

The Enigma Woman: The Death
Sentence of Nellie May Madison
By Kathleen A. Cairns

Front-Page Women Journalists, 1920–1950
By Kathleen A. Cairns

The Cowboy Girl: The Life of Caroline Lockhart
By John Clayton

The Art of the Woman: The Life
and Work of Elisabet Ney
By Emily Fourmy Cutrer

Emily: The Diary of a Hard-Worked Woman
By Emily French
Edited by Janet Lecompte

The Important Things of Life:
Women, Work, and Family in
Sweetwater County, Wyoming,
1880–1929
By Dee Garceau

The Adventures of The Woman Homesteader:
The Life and Letters of Elinore Pruitt Stewart
By Susanne K. George

Flowers in the Snow: The Life of
Isobel Wylie Hutchison, 1889–1982
By Gwyneth Hoyle

Domesticating the West: The Re-creation of the
Nineteenth-Century American Middle Class
By Brenda K. Jackson

Engendered Encounters: Feminism
and Pueblo Cultures, 1879–1934
By Margaret D. Jacobs

Riding Pretty: Rodeo Royalty in the American West
By Renée Laegreid

The Colonel's Lady on the Western Frontier:
The Correspondence of Alice Kirk Grierson
Edited by Shirley A. Leckie

Their Own Frontier: Women Intellectuals
Re-Visioning the American West
Edited and with an introduction by
Shirley A. Leckie and Nancy J. Parezo

A Stranger in Her Native Land:
Alice Fletcher and the American Indians
By Joan Mark

The Blue Tattoo: The Life of Olive Oatman
By Margot Mifflin

So Much to Be Done: Women Settlers on the Mining
and Ranching Frontier, second edition
Edited by Ruth B. Moynihan, Susan Armitage,
and Christiane Fischer Dichamp

Women and Nature: Saving the "Wild" West
By Glenda Riley

The Life of Elaine Goodale Eastman
By Theodore D. Sargent

Give Me Eighty Men: Women and
the Myth of the Fetterman Fight
By Shannon D. Smith

Bright Epoch: Women and
Coeducation in the American West
By Andrea G. Radke-Moss

Moving Out: A Nebraska Woman's Life
By Polly Spence
Edited by Karl Spence Richardson

Eight Women, Two Model Ts,
and the American West
By Joanne Wilke

To order or obtain more information on these
or other University of Nebraska Press titles,
visit www.nebraskapress.unl.edu.